Fruit and Vegetable Processing, Preservation and Value Addition Technology

NIPA® GENX ELECTRONIC RESOURCES & SOLUTIONS P. LTD.
New Delhi-110 034

Fruit and Vegetable Processing, Preservation and Value Addition Technology

Nikita Wadhawan
Head
Department of Food Technology
College of Dairy and Food Technology
Maharana Pratap University of Agriculture and Technology
Udaipur, Uttar Pradesh

Lokesh Gupta
Dean
College of Dairy and Food Technology
Maharana Pratap University of Agriculture and Technology
Udaipur, Uttar Pradesh

Kamla Mahajani
Assistant Professor
Food Science and Nutrition
College of Community and Applied Sciences
Maharana Pratap University of Agriculture and Technology
Udaipur, Uttar Pradesh

Zahara Ali Shams
College of Dairy and Food Technology
Maharana Pratap University of Agriculture and Technology
Udaipur, Uttar Pradesh

NIPA® GENX ELECTRONIC RESOURCES & SOLUTIONS P. LTD.
New Delhi-110 034

NIPA® GENX ELECTRONIC RESOURCES & SOLUTIONS P. LTD.

101,103, Vikas Surya Plaza, CU Block
L.S.C. Market, Pitam Pura, New Delhi-110 034
Ph : +91-11-43860225, Mob.: +91 9717133558, 9540816132
E-mail: newindiapublishingagency@gmail.com
Website: www.nipaersources.com

Print ISBN: 978-93-58873-73-3
ebook ISBN: 978-93-58876-10-9

Composed and Designed by NIPA®.

Preface

This book has been meticulously crafted to align with the course curriculum of VI Deans Committee of ICAR and NEP 2020 regarding post-harvesting technologies, catering specifically to students of Food Technology and Food and Nutrition. Recognizing a longstanding gap in the literature available to students in these fields, we endeavor to address this deficiency by creating this comprehensive resource. Written in a clear and accessible language, we aim to ensure that all students, regardless of their level of expertise, could benefit from its content.

Extensive effort went into the development of this book, involving thorough brainstorming, discussions, and allocation of time and resources. Our primary objective was to compile all relevant sources pertaining to the course, consolidating the latest knowledge strands, elements, and technologies within the subject matter into a single, cohesive guide.

Designed to cater to the needs of students pursuing B. Tech in Food Technology and B.Sc. Food Nutrition and dietetics, this book serves as an indispensable companion throughout their academic journey. We are confident that it will fulfil its intended purpose admirably, providing students with a valuable resource to enhance their understanding and mastery of post-harvesting technologies.

In closing, we express our gratitude to all our seniors and mentors whose guidance and support played an instrumental role in bringing this book to fruition. Their invaluable contributions have helped to ensure that this resource meets the needs and expectations of students in the field of Food Technology and Food and Nutrition.

Authors

Contents

Module 4

List of Tables

List of Figures

Module-1

1

Production and Processing Scenario of Fruits & Vegetables in India and World

India is known as the food basket of the world. It is the second largest producer of fruit till now after China, but its contribution to the Global market of fruit processing is just 1per cent. This is due to the fact that there is deficiency of skilled manpower, poor cold storage facilities, inefficient post-harvest management and minimal technological interventions.

Deficiency of the skilled manpower: In-spite of market size of US dollar 181 billion, expenditure on food and food products being 21per cent of India's GDP and more than 31per cent share of wallet by consumers, there is resultant apprehension about the emergency emerging skill shortages due to mismatch between the demand for specific skills and available supply. Shortage of skilled, Semi- skilled and unskilled workers has emerged as a critical factor impacting the competitiveness of Indian food industry. Efficient man-power is one such critical element to improve to foster the growth. As the food industry advances and adopts various levels of automation the demand of skilled manpower will increase phenomenally. Hence if, the Indian food industry must achieve projected growth target set and become globally competitive, the need to develop and train the human resource efficiently would be essential.

Cold chain facility: To provide integrated cold chain and preservation infrastructure facilities without any break, from the farm gate to the consumer, it covers creation of infrastructure facilities along the entire supply chain viz pre-cooling, weighing, sorting, grading, waxing facilities at farm level, multi product multi temperature cold storage, Controlled Atmosphere(CA) storage, packing facility, Individual Quick Frozen (IQF), blast freezing in distribution hub and refrigerated vans mobile cooling units.

(IQF is a method which does not allow the ice-crystals to form in vegetable cells. Also, since each piece is individually frozen, particles do not cohere and the final product is not frozen into a solid block)

Minimal technological intervention- different food processing technologies are used most effectively keeping in mind the special processing requirements of fruits and vegetables waste issues, nutritional value and consumer concerns. It represents the latest concepts in the fast-changing field of processing of fruits and vegetables.

Challenges in food processing: The modernization during last century resulted in urbanization coupled with modification in life styles and dietary habits. In the same Era the industrial development made it easier to meet the requirements for processed foods. However, consumers are now interested in minimally processed food owing to increase in their awareness to have fruits and vegetables with superior quality, natural integrity and with fewer additives. The food products deteriorate because of physiological ageing, biochemical changes, high respiration rate and high ethylene production. These factors contribute substantially to discoloration, loss of firmness, development of off-flavours, acidification, and microbial spoilage. Simultaneously food processors must use emerging approaches to process perishable commodities along with enhanced nutritional and sensorial quality. So the need is to create efforts to utilize the modern approaches to minimize the processing and deterioration by utilizing techniques of chlorination, ozonation, irradiation, photo synthetization, edible coating, natural preservative use, natural processing, high pressure processing, microwave heating, ohmic heating and hurdle technology. The consequences of the usage of these techniques on shelf stabilization, microbial safety, preservation of organoleptic and nutritional quality and residue avoidance with feasibility and operability of these techniques in modern-day processing is to be determined.

2

Scope of Fruit and Vegetable Processing Industry in India

The term food processing covers an enormous field from simple boiling to the use of irradiation to predict the consequences on the nutritional value of fruits and vegetables. Generally, the food processing methods can be categorized as in intentional processing method (milling, peeling, washing trimming etc.), inevitable processing method (blanching, sterilizing, cooking, shredding, and drying) and accidental (inept processing or system). The nutritional losses associated with fruits and vegetable processing with different processing techniques and conditions employed. For example, the heavily processed products are although safe for human consumption but allied with reduced nutritive value. In this regard access to safe and nutritious food is considered as a basic right of humans as it is one of the ways to prevent various ailments by building better health and enforcing immunity. Escalating help awareness has guided the processes to opt for minimal processing food in recent years.

Food commodities subjected to minimal processing are capturing the market Suring to widespread concept that they might help in meeting the current requirement of vital nutrients. Development of minimally processed commodities especially fruits and vegetables have augmented at both retail and consumer levels during last years. Various organisations WHO, FAO, USDA boost up the utilization of minimally processed perishable commodities including vegetables and fruits to lessen the risk of cardiovascular diseases and cancer. Changes occurring in the eating habits have been quoted due to high anxiety for on time availability of these minimally processed fruits and vegetables. Sales of fresh cut fruits and vegetables continue to burgeon and food industries are persistently expanding primary because of consumers' trend of health realization.

During processing of fruits and vegetables various techniques are employed to improve the quantity and quality of the produce, including blanching, dehydration, salting, and concentration. Most of these Technologies hold adverse consequences on the quality of food, for example blanching is carried out with the major objective to destroy or inactivate harmful enzymes before freezing or drying however it affects the colour, texture, flavour and nutrient value of fruits and vegetables. It is now believed that processing encourages faster detonative changes in fruits and vegetables and ultimately physiological and biochemical changes arise in these commodities even on trivial processing operations. The concepts need to be re- addressed before selecting the appropriate techniques for the processing of fruits and vegetables. There are many modern innovative modules of processing that can be employed to replace thermal processing, thus reducing, and bridging the nutrient losses, resulting in better nutritive quality with greater consumer acceptability.

The meticulousness of fresh cut vegetables and fruits is persistently escalating owing to consumer demand. A distinguishing attribute of minimal processing trend is an integrated approach, where raw material, washing, handling, processing, packaging and delivery to customers must all be properly considered to make shelf-life extension possible. New techniques for maintaining quality and inhibiting undesired microbial growth are on demand at all steps for production and distribution chain.

Food processing recognized as the Sunrise industry provides vital linkages between the two pillars of our economy- manufacturing and agriculture. The food processing industry in India is undergoing a significant transformation. Turnover of dollar 110 billion, it accounts for 35per cent of the Indian food market and has been growing at a better rate of 14per cent in the last few years. This is attributed to the fact of improved policy, environment and increasing thrust on public-private partnership along with improvement to rural infrastructure to leverage strength on the supply and demand frontiers. The government has made commendable efforts to promote investment in the industry by way of channeling resources through various schemes as subsidies and grants. This included development of Integrated Cold Chain (ICC), Mega Food Parks (MFP), Modern Terminal Markets (MTM) and bulk storage facilities as well as modernization of market quality control laboratory and abattoirs. These efforts have placed the food industry on a high growth trajectory as reflected by the increase in food processing sectors.

According to APEDA data, India is the largest producer of vegetables such as ginger and okra and second largest producer of vegetables such as potatoes, onions, cauliflowers, brinjal, and cabbages. While it is the largest producers of fruits such as Bananas, Papayas, and Mangoes. There is a huge opportunity in harnessing the potential of fruits and vegetables processing in India in the form of frozen (IQF), canned, pulp, puree, paste, sauces, snacks, dressings, flakes, dices, dehydration, pickles, juices, slices, chips, jams and jelly.

Andhra Pradesh, Maharashtra, Uttar Pradesh, Madhya Pradesh, Gujarat and Karnataka are the major Fruits producing States, whereas Uttar Pradesh,

West Bengal, Madhya Pradesh, Bihar, Gujarat, Maharashtra and Odisha are the major Vegetables producing States of the country (As per 3rd advance estimates of 2022-2023). The Government of India through the Ministry of Food Processing Industries (MoFPI) is also taking all necessary steps to boost investments in the food processing industry. Under PMKSY, the government has sanctioned 41 Mega Food Parks of which 24 are operational.

As per the APEDA provisional data, exports of processed fruits and vegetables recorded a growth of 9.95 per cent (Apr-Jan 2024), while exports of fresh fruits and vegetables registered a 19.82% growth in comparison to the corresponding months of the previous year. During 2022-23, India exported fresh fruits and vegetables worth $1635.95 Mn which comprised Fresh Fruits worth $770.70 Mn and vegetables worth $865.24 Mn. Exports of processed fruits and vegetables including of pulses touched $2,248.96 Mn which comprised of processed vegetables including of pulses $1,511.14 Mn and processed fruits and juices $737.81 Mn in 2022-23.

With dedicated efforts of the government, installed capacity of fruits and vegetables processing industry has almost tripled from 2.131 MT in 1988 to 9.1 in 2006. Among the states, Maharashtra has largest number of fruits and vegetables (1,048) followed by Tamil Nadu (544), Uttar Pradesh (485) Kerala (445), Karnataka (328), Gujarat (305) and Andhra Pradesh (296). The prominent processed items are fruit Pulp and juices, fruit based ready to serve beverages, change fruits and vegetables, jams, squash, pickles, chutneys and dehydrated vegetables. The major products processed and exported are dried and preserved vegetables, mango pulp, pickle and chutney. The new arrivals in this segment are vegetable curries in retortable pouches, canned mushrooms and mushroom products, dried fruits and vegetables and fruit juice concentrate. Some of the major companies in the segment are as follows:

Table 1: Companies involved in fruit and vegetable processing

Company	**Products**
Capital foods	Frozen foods
Dabur India Ltd.	Jams, pickles, fruit beverages
Godrej F&B	Fruit juices, fresh fruits and vegetables(Retail)
Green Valley	Frozen fruits and vegetables
Hindustan Unilever Ltd	Jams, ketchups, fruit beverages
Mafco	Frozen fruits and vegetables
Mother dairy (Safal)	Frozen processed fruits and vegetables
MTR foods	Pickles and chutneys (dip)
Priya foods	Pickles and fruit juices

Temptation foods	Frozen fruits vegetables and purees
V.P. Baedeker andSons Pvt Ltd.	Spices, pickles, fruits and vegetables,gravy mixes

The FICCI survey identifies that juices and drinks from fruits and vegetables dominate the organized sector, while pickles dominate the unorganized sector. The fruits and vegetable processing industry in India is rather fragmented. Many units are in household and small sector, having low capacities of up to 250 tons per annum. From the year 2000 onwards, the industry has seen a significant growth in ready- to -serve beverages, pulp and fruit juices, dehydrated and frozen fruits and vegetable products, processed mushroom and curried vegetable and units engaged in these segments are export oriented. Preserved vegetables exported by India include dried potatoes, dried mushrooms, preserved onions, dehydrated onion flakes, dehydrated garlic powder, preserved cucumber and gherkins, frozen peas and potatoes, mixed vegetables, frozen and assorted vegetables etc. Major items in processed fruits and vegetables include mango pulp, pickles and chutneys. Pickles exported include chilli, green pickles, mango pickle, tomato and lemon and tamarind concentrate (APEDA 2013). The trends in export of processed fruits and vegetables are given below:

Major destinations for processed fruits and vegetables are as follows:

- Dried and preserved vegetable- Germany, France, Russia, USA and UK
- Mango pulp - Saudi Arabia, Netherlands, Yemen Republic, UAE and Kuwait
- Pickles and chutneys- Russia, USA, Belgium, Netherlands, Spain
- Other processed food and vegetables- United States of America, Saudi Arabia, Netherlands, UK and UAE.

Key Constraints of Food Processing Industry

- Infrastructure in terms of cold storage, washing
- Inadequate quality control and testing infrastructure
- In efficient supply chain and involvement of middle man
- high Transportation and inventory carrying cost
- Affordability cultural and regional preferences of fresh food
- High taxation
- High packaging cost

India is endowed with a remarkably heterogeneous area characterized by a great diversity of agro-climatic zones. It allows to production a variety of horticultural crops such as fruits, vegetables, flowers, spices, plantation crops, root and tuber crops, and medicinal and aromatic crops. India's diverse climate ensures availability of all varieties of fresh fruits & vegetables. It ranks second in fruits and vegetables production in the world, after China. As per National Horticulture Database published by National Horticulture Board, during 2015-16, India produced 90.2 million metric tonnes of fruits and 169.1 million metric tonnes of vegetables. The area under cultivation of fruits stood at 6.3 million hectares while vegetables were cultivated at 10.1 million hectares. India is the largest producer of ginger and okra amongst vegetables

And ranks second in production of potatoes, onions, cauliflowers, brinjal, Cabbages, etc. Amongst fruits, the country ranks first in production of Bananas (25.7%), Papayas (43.6%) and Mangoes (including mangosteens and guavas) (40.4%). The vast production base offers India tremendous export opportunities. During 2018-19, India exported fruits and vegetables worth Rs. 10236.93 crores/ 1,469.33 USD Millions which comprised of fruits worth Rs. 4817.35 crores/ 692.01 USD Millions and vegetables worth Rs. 5419.48 crores/ 777.25 USD Millions. Grapes, Pomegranates, Mangoes, Bananas, Oranges account for larger portion of fruits exported from the country while Onions, Mixed Vegetables, Potatoes, Tomatoes, and Green Chilly contribute largely to the vegetable export basket. The major destinations for Indian fruits and vegetables are Bangladesh, UAE, Netherland, Nepal, Malaysia, UK, Sri Lanka, Oman and Qatar. Though India's share in the global market is still nearly 1% only, there is increasing acceptance of horticulture produce from the country.

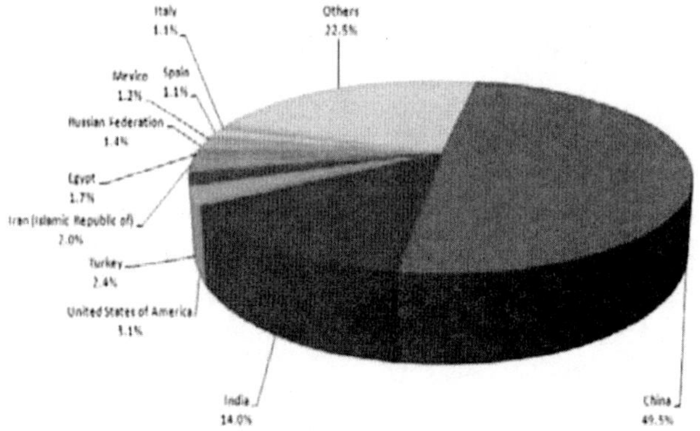

Figure 1: Major vegetables producing countries

- India is the second largest producer of the fruits and vegetables in the world after China. The country has diverse topography and climate, which ensures availability of fruits and vegetables in every season. During 2021-22, India produced 204.61 million metric tonnes of vegetables and 107.10 million metric tonnes of fruits. The area under cultivation of fruits stood at 7.09 million hectares in 2021-22, while vegetables were cultivated at 11.28 million hectares. In addition, India leads the world in the production of ginger and okra, along with banana, papaya, mango and guava.

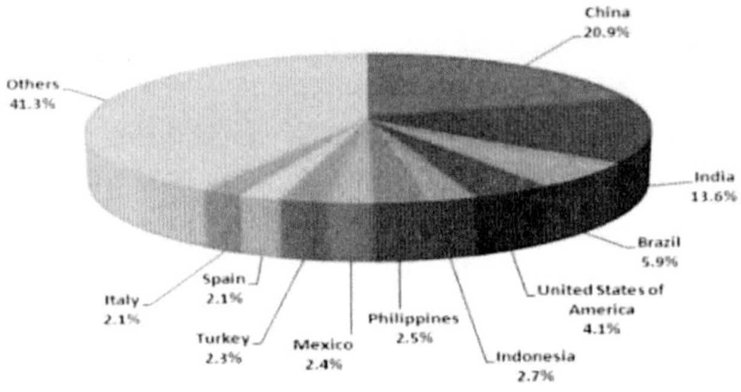

Figure 2: Major fruit producing countries

- Production and export of Indian fruits and vegetables have increased in recent years. During 2021-22, India exported fresh fruits and vegetables worth US$ 1.527.60 million, which comprised fruits worth US$ 750.7 million and vegetables worth US$ 767.01 million. India mostly exports to its neighbors, namely the UAE, Bangladesh, Pakistan, Saudi Arabia, Sri Lanka and Nepal. India's global market share is still only about 1%, but its horticulture products are becoming increasingly popular due to advancements in cold chain infrastructure, research, contemporary post-harvest technologies, supportive governmental policies and quality control procedures.

Production of Fruits and Vegetables in India. According to the United Nations' Food and Agriculture Organization (FAO), India was the world's second-largest producer of fruits and vegetables. India's food industry is large, and the biggest fruit- producing states are Andhra Pradesh, Maharashtra, Madhya Pradesh, Uttar Pradesh, Tamil Nadu, Karnataka and Gujarat. The major vegetable-producing states are Uttar Pradesh, Madhya Pradesh, West Bengal, Bihar, Gujarat, Odisha, and Maharashtra.

Fruit production in India increased to 107.10 million tonnes in 2021-22 from 97.97 million tonnes in 2018-19, at a CAGR of 3.0 per cent. Further, the vegetable market production expanded by 3.8 per cent CAGR to 204.61 million tonnes during this period.

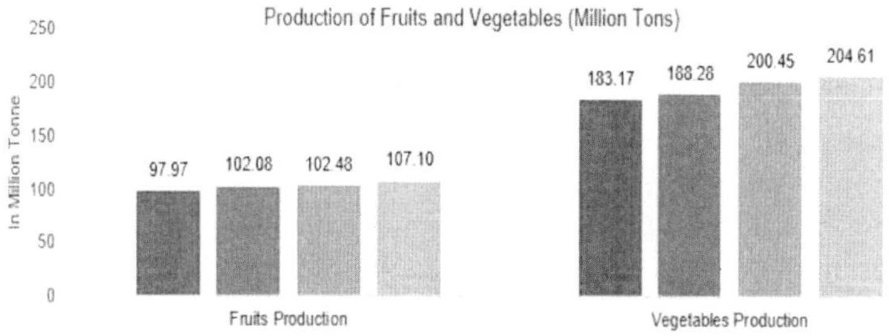

Figure 3: Production of Fruits and Vegetables

Fruits and Vegetable Exports from India: The large production base provides India significant export potential. It exported fresh produce worth Rs. 11,412.50 crore (US$ 1,527.60 million) over 2021-22, with fruits valued at Rs. 5,593 crore (US$ 750.7 million) and vegetables at Rs. 5,745.54 crore (US$ 767.01 million). During this period, processed fruits and vegetables (including pulses) were exported for a total of Rs. 12,858.66 crore (US$ 1,724.88 million), of which Rs. 8,308.04 crore (US$ 1,114.19 million) and Rs. 4,550.62 crore (US$ 610.69 million) were processed vegetables, including pulses. The major fruits exported are grapes, pomegranates, mangoes, bananas and oranges, while the major vegetables exported are onions, mixed vegetables, potatoes, tomatoes and green chilies.

According to preliminary data from the Directorate General of Commercial Intelligence and Statistics, processed fruits and vegetables saw considerable growth of 51 per cent between April and July 2022 compared to the same period last year, while fresh fruits and vegetables saw an increase of 3.79 per cent. Fresh fruit and vegetable exports totaled US$ 498 million during this period, rising to US$ 517 million in the comparable months of the current fiscal year. In the first four months of the current fiscal year, processed fruits, and vegetables exports more than doubled to US$ 665 million from US$ 441 million in the equivalent months of the previous year.

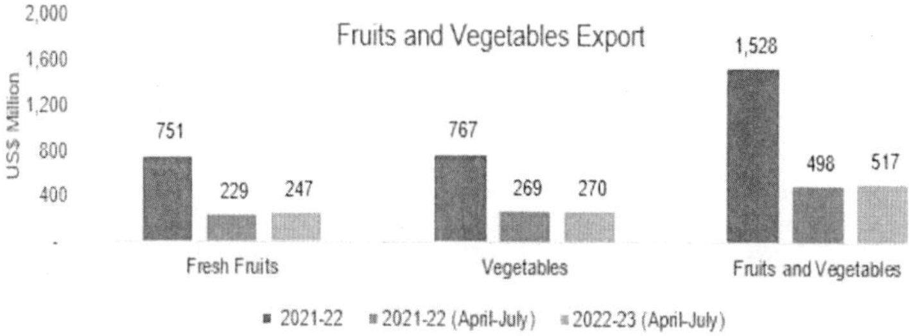

Figure 4: Fruits and Vegetables Export

India's food processing sector is one of the largest in the world and has gained prominence in the recent years. The availability of resources and encouraging government policies have contributed significantly to the industry's expansion. Fruits and vegetables are the key sub-segment of this industry. For 2022-23, the government has fixed an export target of US$ 23.56 billion for the agricultural and processed food industry, 40 per cent of which has already been achieved in the first four months of this year. South Asian & Middle Eastern countries have been the major destinations for India's fruits and vegetables. However, it has huge potential to increase its presence in untapped markets soon.

3

Overview of Principles and Preservation Methods of Fruits and Vegetables

Botanically fruits are ripened ovary. It is a plant part that may or may not be purely derived from ovary of the flowers may be fleshy or pulpy in character, often juicy, predominantly sweet. With somewhat acidic taste, fragrance, aromatic flavours, fruits are generally not consumed during principal part of meals. Whereas vegetable is plant part which is less sweet, generally non acidic and consumed either cooked or raw during the principal part of our main meals. Vegetables may be fruit, root, stem flower, or leaf part of a plant or their modification.

Fruits and vegetables are valued in in human diet as an important source of protective nutrients like Minerals and vitamins. Apart from their nutritive value they add variety to the diet and improve acceptability of meal due to presence of different shades of colours, flavours, and texture. However, the major role in food preservation is played by their moisture content. For the same reason they are classified as

1. **Perishable**: Those which deteriorate quickly after harvesting such as tomatoes, mangoes, papaya, peaches, plums, juicy fruits and vegetables. They have high degree of moisture content and are highly susceptible to spoilage.
2. **Semi perishable**: These have less moisture content as compared to perishable one's example 30 carrots, peace, green beans, apple etc. These foods are containing certain natural inhibitors to spoilage.
3. **Non-perishable**: The fruits and vegetables with very low. These are not easily susceptible to spoilage by microorganisms and enzymes. Fruits and vegetables do not form part of this group.

Still the fruits and vegetables are cheaper and better food sources which significantly contribute to human well-being. If they can be supplied in fresh or preserved form throughout the year for human consumption the national picture will improve greatly.

Food preservation can be defined as the science which deals with the methods of prevention of decay or spoilage of food, thus allowing it to be stored in a fit condition for future use. It may also be defined as the state in which any food may be retained over period of time without being contaminated by the pathogenic organisms or chemical and also without losing the optimum qualities of colour, texture, flavour and nutritive value. Preservation of food may also reduce its preparation and cooking time and becomes advantages of adding more convenience to it. Thus, the food preservation advantages can be abridged as:

1. It adds variety to the food
2. Increases shelf life of the food
3. Helps to increase the food supply
4. Decreases the wastage
5. It also decreases dietary inadequacies

Food spoilage is defined as any decay or undesirable decomposition of constituent by excessive growth of microorganism or by the other physical and chemical causes. Food becomes unfit for human consumption if maximum limit of presence of microorganism crosses the prescribed one. The major cause of food deterioration includes:

1. Growth and activities of micro-organisms., principally bacteria, yeast and mould
2. Activities of food and science and chemical reactions within the food itself.
3. Infestation by insect, parasites, and rodents.
4. Either gain or loss of moisture.
5. Reaction with oxygen.
6. Effect of light.
7. Physical stress or abuse.
8. Time period.

The major causes may be grouped as

1. Biological- growth of bacteria, fungi, mould, activities of enzymes, insects, rodents and parasite.
2. Chemical- reaction with oxygen, chemical constituent, and light.
3. Physical- temperature, physical stress, and abuse.

All these factors do not operate in isolation. Bacteria, insects, and light for example can all be operating simultaneously to spore food in the field or in in warehouse. Similarly, heat, moisture and air simultaneously affect the multiplication and activities of bacteria as well as chemical activities of food.

Biological Changes

- Growth and activity of microorganisms: Most significant changes in food are due to biological factors specially microorganisms. Microorganisms are present in air, water on most of the foods and on our hands etc. They are so small that they can only be seen through microscope. Types of microorganism which cause spoilage are:

  1. **Bacteria:** are found everywhere minute unicellular organisms. For growth they require a proper kind of food material, heat, pH, moisture and oxygen
     - Are the most common causes of spoilage
     - Are much more difficult to be killed than mould and yeast.
     - In active form (vegetative stage) destroyed at boiling temperature
     - Resting form (spore stage) are difficult to destroy and requires application of heat for a long time. Boiling 100 degree centigrade for 6 hours for 121 degrees centigrade for 30 minutes under 10 LBS pressure.
     - Due to acid and can be easily killed in acidic pH at the temperature of boiling water (100 degree centigrade).
     - For food with low acid content, processed at Higher temperature (116 degree centigrade) in a steam pressure to kill bacteria at for faster rate
  2. **Yeast:** Microscopic unicellular organisms which are non-motile, round or oval are normally called as yeast which multiply or reproduce by a characteristic process of "budding". The bud when fully mature, read away from the mother cell and become independent and repeat the process of multiplication.

     Yeast require less moisture and acidic pH to grow and do not grow in alkaline medium. They grow well in light sugar solution and acidic medium.

     Are very useful and making bread, beer, wine, vinegar and many e other fermented products. Yeast causes fermentation of Sugars and convert them into alcohol and carbon dioxide. This property is important in making alcoholic beverages and in baking

     sugar + yeast= alcohol+ CO2

3. **Mould:** Moulding of fruit is quite common. A slice of white bread or a piece of orange left for a time the becomes covered with a whitish or Grace ordinary matter. This is mould. They grow in a network of hairlike fibers called mycelia and send a fruiting body that yield spores. They thrive best in close, damp and dark situation and require adequate supply of warmth, moisture and air for growth. The aerobic and require O2 and less available moisture and can grow well at 25-30 degree centigrade. They prefer sugar containing substances like jam, jelly preserve others sweet based product. They can also grow at wide range of pH (2-8.5) but majority grow well at acidic pH Therefore, they grow on pickles juices etc.
4. **Action of enzymes**: Enzymes produced by living things are compounds which catalyze chemical reactions. Found in all Fresh Food- plants and animals. Action is important for ripening of fruits and vegetables up to certain stages enzyme action may be desirable but its continuous action ultimately render the food incredible by bringing undesirable changes in food tissues.
   - Darkening of cut surfaces
   - Ermentation of soft spots
   - Development of off-flavours

   Enzymes convert starch- sugar, protein- amino acids, pectin- pectic acid and therefore change the constituent of food. Some vegetables and fruits john Brown when damaged or when their cut surfaces are exposed to air due to presence of enzymes as phenolase, peroxidases and polyphenol oxidases. Natural tendering or aging of meat is due to action of enzyme. It can be in activated by heat, cold drying, addition of inhibition chemicals and radiation. It can also be controlled by regulating temperature and exploring moisture and air. Enzymes can act L/W (15.6-65.6 degree centigrade). Optimum temperature 37-degree centigrade drink protein related by heat. Activated by heating at temperature between 70 degrees centigrade to 93.3 degree centigrade.
5. **Insect parasites and rodents:** Insects, worms, bugs, fruit flies damage food stuffs such as cereal grain and fruit and vegetable and rendered them unfit for human consumption the injuries caused by them serve as Pathways by which microorganisms reach the inner tissue and cause spoilage in food.
6. **Purely chemical action:** These are those by which are not catalysed by enzymes of tissues or microorganisms example rancidity of fats.

Atmospheric oxygen and even sunlight cause certain foods to undergo various undesirable changes. Oxidation results in production of off flavour as well as loss of ascorbic acid. Reactions are called lipid oxidation/ non enzymatic reactions etc... Maillard reaction is one of main major causes of deterioration which occurs during storage of dry food. The deterioration includes of flavour, colour development and surface biting. Banana, tomato, lemon and squash shall be stored at about 10 degrees centigrade for retaining their quality.

7. **Physical and mechanical damage:** All physical changes caused by freezing, burning, drawing selection of packaging material etc. During transportation and storage of food also cause spoilage. Sometimes fruits are bruised or scratched during harvesting and handling. If precautions are not taken, the injured spot become the points of entrance of microorganisms which cause spoilage, example stem and root in mango, green mould in in citrus fruits and blue mould in Apple.

   Removal of deceased or damaged and scratched fruits during grading or post-harvest treatment of fruits with fungicide like thiabendazole etc. Are highly Useful in reducing mechanical damage to fruits.

8. **Storage conditions**: Temperature, oxygen, lights, duration of storage etc are important factors that type of microbial growth and spoilage of food during storage.

9. **Temperature**: Brain denaturation and destruction of vegetable. Even at refrigerating temperature certain fruit and vegetable suffer discoloration and change in texture.

   - Freezing of milk: Emulsion broke and fat separates, denatures milk protein
   - Chill Injury
   - Moisture: excessive Pick up aur loss causes substantial deterioration.
   - Oxygen: Atmospheric oxygen brings about undesirable changes in food such as destruction of food colour, flavour and Vitamin a & c.
   - Light: All other food deteriorative factors, longest storage time, greater deterioration except for fermented food which improve up to a point with aging.

Principles of Food Preservation

In the preservation of food by different methods, following principles are involved

1. Prevention or delay of microbial decomposition
   a) By keeping out microorganism (asepsis): Its means preventing entry of microorganisms that is maintaining general cleanliness while picking, grading, packing and transportation by carefully washing and wiping the fruits and vegetables. Superior quality prepared products can be obtained.
   b) By removal of microorganism (filtration): In this method juices are clarified by settling or by using ordinary filters and then passed through special filters which can retain yeast and bacteria. Various types of germ proof filters are used for this purpose. Recently in USA and Germany this method is used for preserving apple and grapes juices, soft drinks, and wine.
   c) Buy hindering the growth and activity of microorganism: By low temperature, drying and aerobic condition or by using chemicals or antibiotics.
   d) By killing the microorganism example by heat or radiation.
2. Prevention or delay of self- decomposition of the food
   a) By destruction or inactivation of enzymes, for example by blanching.
   b) Prevention or delay of chemical reactions, for example prevention of oxidation by means of an antioxidant.

      This method employs not only destruction of enzymes but also elimination. Self-decomposition can sometimes also be prevented by the action of purely chemical reactions.
3. Prevention of damage by insects, animals and mechanical causes Foods if not packaged or stored properly are attacked by insects and worms etc. which cause damage to the food product and make them unfit for consumption. Hence, to retain the natural skin and aroma of a product, it is necessary to preserve it soon after preparation, without allowing it to stand for any length of time. This damage can be prevented by use of chemicals and fungicides.

Table 2: Principles of food preservation

S. No.	To be controlled	Heat	Cold	Chemicals	a_w	Mechanical
1	Microorganisms	Prevent growth	Reduces growth rate	Retard growth	Do not grow below aw of 0.6	Reduces number
2	Enzymes	Destroyed by heat activity	Decreases reaction rate	Modify activity	Alters rate of enzyme activity	Increases enzyme substrate complex formulation
3	Chemical reactions	Increases chemical rate	Reduces reaction rate	May inhibit oractivate	Alters rate of reaction	Not applicable
4	Physicalstructure	Increases effects	Decreases effects	May modify structure	High aw maycause caking	Can destroy structure

Module-2

1

Supply Chain of Fresh Fruits and Vegetables

India produces 146 million tons of vegetables (NHB report, 2011) and the total area under vegetable cultivation is around 8.5 million ha. Potato, tomato, onion, cabbage and brinjal account for around 60 per cent of the total vegetable production in terms of quantity. India's annual fruit production is 75 million tons (NHB report, 2011) and the total area under fruit cultivation is around 6.3 million ha. The major Indian fruits are mango, banana, citrus fruits, apple, guava, papaya, pineapple, and grapes. Fresh fruits and vegetables play a vital role in Indian context and their marketing works as an influencing force in the economy. Though retailing of fresh fruit and vegetables is a low margin business, the market potential is very large in a country like India. This factor has attracted many corporates into this sector. Marketing of agricultural produce is different and more challenging than many industrial products because of the perishability, seasonality, and bulkiness. The very nature of small size of land holdings by the farmers, varied climatic conditions, production spread over wide geographical area, mainly in remote villages, diversified consumption patterns/habits of the Indian consumers and poor Supply Chain (SC) infrastructure makes marketing for fruits and vegetables more complicated. At the same time, Indian consumers demand fresh fruits and vegetables. Thus, Management plays a crucial role in marketing fruits and vegetables. Supply Chain efficiency not only helps in increased production and per capita consumption, but also contributes to economic development of the country. As a result, Supply Chain Management (SCM) throws both challenges and opportunities in marketing of fruits and vegetables. Efficient SCM in marketing, not only increases the profitability and efficiency of retailers, but also adds value to different stakeholders like cultivators (farmers), consolidators and consumers. Traditional Retail Model of fruit and vegetables is predominantly followed in unorganized retail sector. Players involved in this model are commission agents, auctioneers, wholesalers, traditional retailer of all type of formats family run 'mom and pop' stores, roadside shops, pavement shops and cart vendors apart from farmers and customers.

Vegetable Retail Model

Presently the fresh vegetable retailing in India is carried by two distinctly different types of business groups. They are classified as 'organized' and 'traditional' organizations. Now, the vast majority urban market and rural market and are served by traditional fresh vegetable and grocery retailers. Indian food retail consists of staple commodities comprising grains, pulses and fresh vegetables. Indian food retail, especially fresh vegetable retailing is witnessing a rapidly growth in India's organized retail sectors. The traditionally retailing of fresh vegetables in India is not very much organized which amounts to 97 per cent of the total market extremely localized and highly fragmented with large number of intermediaries. The intermediaries between the customers and farmers are traditional retailers with different outlet formats – mom and pop shop, non-permanent shops in the market, pavement vendors, road side vendors and push cart fresh vegetable sellers, wholesale traders, commission agents and auctioneers. The farmers themselves sell their products directly to the end consumers as sellers in local markets, regulated and unregulated 'Farmers markets', sell to intermediaries - agents and organized retailers. The market place is near their farm land and customers accessing the market are in a closer distance. Farmers selling fresh vegetables directly to the customer are amount to very small fraction by volume. Farmers sell bulk of their produces to agents and auctioneers. The agents buy even small quantities of produces from farmers and transfer it to wholesalers directly or through another agent. The auctioneers are those who have entered buying contract with farmers for whole or partial quantity of the produces and sell the produce to an agent or a wholesaler. Auctioneers also transfer the fresh vegetables to wholesalers directly or through another agent. Wholesalers of fresh vegetables sell to retailers, both traditional and organized retailers and to customers, who buy in large quantity. Cart vendors, a type of tradition retailers, buy in fresh vegetables from wholesalers or organized retailers, sell to customers in mobile carts and deliver to customers at customer's door steps. Whole sale market is a vital link in fresh vegetable supply chain. Both the traditional and organized retailers are impartibly dependent on whole sale market with different propositions.

Food Mileage

Food Mileage is an indicator to evaluate impact on economic, social, and ecological system and it associates the quality food availability, foods wastage and disposal. The term 'Food Miles' (or 'Food Kilometers') refers to the distance food travels from the location where it is grown to the location where it is consumed, or in other words, the distance food travels from farm to plate. Food miles do not refer the input material, effort, efficiency, or energy to the crop yield. Food miles are a way of attempting to measure how far food has travelled to reach consumer. That includes the journey from farm to processor, then from processor to retailer and finally from retailer to consumer. Shorter food miles are an indicator of near sourcing and longer food miles of fresh vegetables is an indicator for agricultural outsource. 'Food miles' is a factor to understand inefficiency of food supply chain. In economical or business perspective, every food mile is cost. Transportation cost is directly propositional to food miles. Every mile addition in transport is addition in cost of the goods. Customer pays for it. Fresh vegetables travel for miles, the less freshness. Means customers pay for fresh vegetables which have less initial nutritional value. Alternatively, to retain freshness, conditioning is required while transporting. Fresh vegetables es with less mileage are afresh, original taste, initial ingredients retained and more palatable. Less food miles create more sense of closeness and trust. Ecologically, 'food mileage' is a convenient indicator of sustainability and sustainable Development; less food miles indicate more sustainability. Reducing food miles is reduction of emissions. Shorter distance travels lead to reduced usage of fossil fuels and thus, conservation. Minimum food travel: minimum pollution, environmental degradation, and Global warming. Studies estimate that processed food in the United States travels over 2080 kilometers (1300 miles), and fresh produce travels over 2400 kilometers (1500 miles), before being consumed. In UK, twenty percent of food (by weight) moves more than 200 kilometers.

Marketing of Fruits and Vegetables is quite complex and risky due to the perishable nature of the produce, seasonal production, and bulkiness. The spectrum of prices from producer to consumer, which is an outcome of demand and supply of transactions between various intermediaries at different levels in the marketing system, is also unique for fruits. Moreover, the marketing arrangements at different stages also play an important role in price levels at various stages viz. from farm gate to the ultimate user. These features make the marketing system of fruits to differ from other agricultural commodities, particularly in providing time, form, and space utilities. While the market infrastructure is better developed for food grains, fruits and vegetables markets are not that well developed and markets are congested and unhygienic. Different marketing channels include • Producer-trader-wholesaler-retailer- consumer. • Producer-trader-retailer-consumer. • Producer- trader-consumer. • Producer-consumer. An efficient marketing system can • Reduce post-harvest losses. • Enhances farmers 'realization. • Reduce consumer price. • Promote grading and food safety practices. • Induce demand-driven production. • Enable higher value addition. • Facilitate export.

Newly Developed Supply Chain Models for Fruits and Vegetables

All previously mentioned models were depicted either processing based or retail-based supply chain from farm gate to retail outlets, whereas Figure 3.9 and 3.10 shows the fruits and vegetables supply chain model respectively by combining both processing (value added fruits and vegetables) and raw consumption (without value addition). These models were developed as new by combining all the above models and validated the same by presenting various conferences and publishing different journals. These supply chain models include five different supply chain 67 stages such as farm gate, wholesale (traders), cold storage, processing, and retail under two types of distribution flow namely, raw consumption and processing. Both types of flow distributions were started from farm gate, ended with retailing, and supported with traders and cold storage. Raw consumption flow reached to the retail stage without value addition whereas processing flow reached with value addition.

Primary processing industry was defined as extracting pulp from the fruits whereas, secondary processing industry defined as producing end products like jam, jelly, squash etc. Primary processing is applicable only for fruits and not for vegetables since there is no possibility for extracting pulp from vegetables except tomato. Hence Figure 3.9 shows the supply chain model for fruits along with primary processing stage, whereas Figure 3.10 shows the supply chain model for vegetables without primary stage.

Major Issues in supply chain of Fruits and Vegetables

1. **Cold Chain Management issues:** There are various issues related to cold chain in India, such as lack of cold chain facilities, inadequate capacity of cold chain, lack of cold chain network. Due to this concern, it has become difficult for the farmers and businessman to do their business effectively and get proper remuneration for their produce.
2. **Infrastructure issues:** Supply chain infrastructure plays an important role in the F&V sector. Proper and adequate infrastructure helps farmers and agri- businessman to run their business successfully and helps to deliver the goods in the right time with right condition. In India, Infrastructure in the main impediment in the supply chain of agricultural products which leads to high number of losses.
3. **Packaging issues:** Packaging is very important for F&V as they are highly perishable goods and it needs proper packaging for the handling of these fresh produce. Without proper packaging it is very difficult to maintain their shelf life. Cost is very important factor for this issue. High cost of packaging material makes difficult for the farmers to do proper packaging of their goods.

4. **Technological issues:** The technology is surrounded by many technical issues, such as advancement issues, inefficient technology, obsolete techniques, and old machineries. Due to these concerns, it has become difficult for the farmers and agri businessman to use an appropriate technologies and techniques to reduce the post-harvest losses and time in operational activities.
5. **Farmers' Knowledge and Awareness:** Farmers in India have very poor knowledge regarding the use of latest technologies, techniques etc. to work effective and efficient. They have very poor knowledge regarding the management of post-harvest produce, quality of seed etc. Without proper knowledge and awareness level of the farmer the supply chain of F&V cannot be efficient, because farmers are the main source of the fresh agri produce.
6. **Quality issues:** Quality is a very important factor in food industry/ sector, because it directly relates to the health of the people. It is very important for supply chain to deliver the fresh goods in a timely manner and in a proper quality to the customer. Proper supply chain helps to maintains the shelf life of produce and prevent from deteriorating the quality. Quality has a strong impact on the supply chain, so it leads to efficiency and less rejection by the customer. In India, there is a lack of quality standards to meet international quality for export, Poor hygiene and safety standards, high quality degradation etc.
7. **Financial issues:** Income of farmers is very low in the state. They don't get proper income for their agricultural produce and maximum amount of share eat up by the intermediaries. There are lack of transparencies in pricing at all Mandis in India due to which farmer don't get right prices for their efforts and fresh produce.
8. **Transportation issues:** Transportation plays a very important role in the supply chain. Without proper transportation the goods can't be delivered to the customer in a right time and in a right quality. It plays even a more important role in Perishable food like Fruits and Vegetables because of short shelf life, high perishable, required controlled temperature. Transportation related challenges are very high in the India because of unavailability of well transportation mode, high cost of transportation, lack of temperature-controlled vehicle for the movement of goods etc.
9. **Block Chain:** Blockchain is an example of a Distributed Ledger Technology (DLT). The key thing about this technology is that it is a 'distributed', as opposed to 'centralised', system. Therefore, rather than having one central system that is owned by one body, a blockchain

system means that all the data is recorded in many places, each of which is owned by the actors within the blockchain. For example, rather than having a single central system that is controlled by the government or a brand, the model is distributed so that all of the actors, suppliers and customers accessing the supply chain have their own copies of the records, but only if they are allowed. So, a blockchain is open, easily allowing anyone on, but private in that membership is controlled.

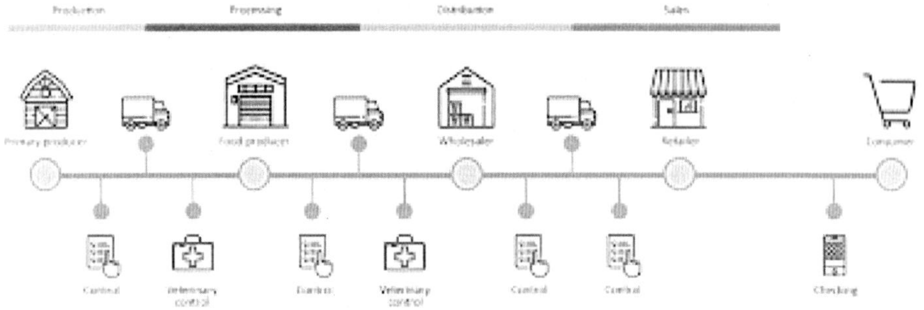

Why is blockchain useful in the food industry?

Blockchain can be used in the food industry because, fundamentally, it is about helping to manage the food system, which is itself a chain, or set of linked chains. There are many actors in that chain; there are suppliers who are providing the food, and there may be aggregators and retailers and people who are ultimately selling the food, but it is all about the supply chain. Blockchain, as the name might imply, is very much aimed at helping to manage a chain of actions and consequential transactions. Although not 100 percent tamper proof, the technology that underpins blockchain is extremely secure and lends itself to the food system.

How Does Blockchain Work?

If every single transaction is being shared with everybody across the chain, the actors on the chain become self-regulating and able to manage their own chain. So, if one node is attacked and compromised from a security point of view, all the other nodes are still there. It is much more resilient, and everybody can see that the chain has now been compromised and, therefore, something has gone wrong, allowing the relevant bodies to act.

Rather than having one central system that is owned by one body, a blockchain system means that all of the data is recorded in many places

If tampering did occur, this would create a different block – one that you can see has visibly changed. It is designed so that every relevant body would be able to see that A had become B, at a specific point. Blockchain is different to something like Bitcoin, which is using similar technology. Bitcoin is open to anybody – anyone in the world can decide they want to buy some Bitcoins and see all that data about the Bitcoin. On the other hand, whilst it is sharing visibility and is transparent, blockchain is only transparent to those people who have joined that blockchain. So, when you join you effectively have a contract to say: "I'm joining the chain, I'm becoming a member of the group." It is a bit like having a cooperative. I am going to become a member of the group and therefore I see everything that I am allowed to see within the group, but it cannot be shared with the general public – it is not that open and transparent – but it is open and transparent within the controlled group.

Who Regulates the Chain?

Control of the whole chain is where it becomes slightly more complicated and interesting. The work that we have done in the Food Standards Agency (FSA) explored the application of blockchain in two areas. One was in relation to cattle going into an abattoir and the need for the data about the cattle to flow into the abattoir and the need for the data about the carcass to flow back. So, there is a use-case with lots of different actors and that worked well. Everybody was happy, it was easy to deploy, and it was relatively cheap technology-wise. Everybody got value from it...what is not to like?

A blockchain is open, easily allowing anyone on, but private in that membership is controlled. But we had the same question...who – which one of the actors – should own this chain? Is there a third party that coordinates this? Or, again, is it a little bit like a cooperative in how it operates? And that sort of operating model question is one of the outstanding questions that we are now faced with. The FSA kicked off that pilot, paid for it, and commissioned the development of the system, but the FSA then said: 'We found it very valuable, we get access to the data that we need and we are using data that has already been entered. So, it is very cost effective. But the FSA's role is not to establish blockchains for every single step in the UK food supply chain.'

So that question around operating is one we are still looking for answers for. We are talking with others, such as the US Food and Drugs Administration (FDA) and HMRC to find that answer. The other conclusion that we can draw is around trust. For a body to join a chain, all parties involved should be able to trust that body. The trust is then demonstrated and maintained through the performance of that body on the blockchain.

Can we Trust Blockchain?

There are different levels of trust. One is the original entering of the data, and the other is agreeing that the data is going to be shared, and that is where some people say, 'I am not sure about that. I don't want you to see this data. It is commercially sensitive' for example. And that is not only people saying they do not want to share data with me, the regulator, it may be a supplier that does not want to share data with a retailer, or a retailer that does not want to share data with another competitor. We are now looking at something called 'data trusts', which is a way of sharing the data in a very controlled way – in other words it has limitations as to who can see and do what with the information.

Then, when considering the actual entry of data, one thing to consider is that someone might ask: 'The data on the blockchain is only as good as the data that is being entered. Who is saying that that carcass did come from that animal? And who is saying yes that carcass is fit to eat, or has these problems, or it is this weight? The blockchain cannot physically know that that carcass is the one it is said to be, so how can I be confident in the data on the blockchain? 'Well, two things: one is the fact that it is no worse than it is today. Your existing IT systems would be in the same place. The second is that there are other pieces of technology such as IoT sensors, that allow you to do that final step to connect the blockchains to something that says: 'I can physically guarantee that that is that carcass.'

Do you have any examples of successful blockchain implementation?

The FSA carried out another pilot focused on pork. The aim was to provide confidence in provenance data in relation to pork being exported to China. Similar to the beef pilot, it was easy to deploy this trial, it was quick and there were no technical problems. But, again, it raised the question around the overall operating model. The third pilot we have been working on is with HMRC, and this is two regulators working together for the import of wine from Australia. If the Australians are exporting wine to the UK, the data that they provide on their blockchain about the wine is used by UK business, and by different regulators, us and HMRC, but we all do it once on one system. This is more efficient for each of us, and there is less confusion as to what exactly is the correct data.

The regulator should not have to be there all the time monitoring the data that flows through, but, as and when required, have rights to access data

There are also a few other examples; Walmart is one of the biggest examples – the supermarket is using blockchain to join up the teams within an organization. Rather than using blockchain across organizations, it is implementing it internally to join up its own supply chains within the company.

This allows the company to understand what is going on in its supply chain at speed. As it is an internal operation, it also negates the issue of 'who is the control of the chain. 'A helpful use case of blockchain I can envision for a business would be for recalls, as it will enable a company to trace back to the source and follow the food on its journey in real-time. This could mean that recalls could occur in near to real time too, minimizing the amount of product recalled, and sorting the problem quickly.

Is there a Blockchain Alternative?

The alternative to blockchain are traditional style IT systems. You could say: 'I can just carry on my existing IT system' and it does a fair amount of what blockchain would do. It is not to say it is a clear competition between these types of technologies – it is more complex than that. I think much of what we are seeing in our learning is that it is not about the technology, but it is more about operating models and trust. There are different levels to the technology and there are various types at the underpinning technology level, it is more how these operating models are constructed, and how they are used. We want to ensure several things. One is that we, and industry, do not get locked in. We do not want there to be a single supplier of this technology. Government should do its best to try to avoid that from happening. The other thing is that the regulator is another actor in this space, so if blockchains are established, the regulator would like to ensure that it is an actual secure blockchain. The regulator should not have to be there all the time monitoring the data that flows through, but, as and when required, have rights to access data.

Another thing to consider is data standards. A blockchain, or any other technology that will join up the supply chain, requires standards to allow each chain, each element, to be able to talk to each other. That is critical and something that regulators must work with industry to focus upon.

What is the Future of Technology and Food Integrity?

There are many promising emerging technologies, and we will continue to monitor them as they appear; we will continue to undertake our own research and explore what the various technologies might be able to offer. A priority for the FSA at the moment is in relation to analytics – using data for predictive analytics. I think it will be important for the food system, and although it is already doing a lot, more can be done to really understand what is going on, analyze trends, see things going wrong and better predict the future. And again, as the regulator, it would be beneficial to be able to see what the risks are that might be impacting consumers, either today or in the distant future.

2

Primary Processing and Pack House Handling of Fruits and Vegetables

The types of operation carried out vary with different commodities and market requirements. Produce that are destined for export or supermarket outlets are often subjected to elaborate operations compared to local markets. Some of the operations are as follows:

A. **Sorting and trimming:** Freshly harvested fruits and vegetables are sorted for uniformity in size, shape, and varietal characteristics. Damaged, discoloured, and decayed parts are removed to make the produce more attractive and prevent infection from the diseased parts.

B. **Washing:** It is necessary to remove extraneous materials from the field such as dirt, chemicals and latex. This is usually done before storage or immediate retailing of the produce. To reduce the incident of decay, chlorine is often added to the wash water.

C. **Drying:** It is done to remove excessive moisture from the surface of the produce. Excessive drying should be avoided to prevent wilting, shrinking and water loss.

D. **Waxing**: is done on certain types of fruits and vegetables such as ginger, tomato, citrus and melons to reduce water loss, thereby reducing shriveling. In addition to that, the application of edible wax will enhance the appearance of the produce.

E. **Curing**: Injured and bruised surfaces of root, rhizome and tuberous crops are allowed to heal by holding them at ambient temperatures for a few days. Curing initiates the formation of periderm layers at wound areas, thereby reducing moisture loss and microbial infection. Sufficiently cured vegetables can be stored for a longer period.

F. **Chemical Treatments:** Fungicides and growth regulators are commonly used to reduce decay and undesirable growth respectively. The use of chemicals should be closely supervised and within the recommended levels for human consumption.

G. **Grading:** Fresh fruits and vegetables are classified according to a set of recognized criteria of quality and size, with each bearing an accepted name and size grouping.

H. **Packaging:** Produce is packed in suitable containers to provide protection against mechanical and biological damages during transportation and subsequent handling operations. Packaging materials should be of accepted standards regarding strength, ruggedness, and resistance to pressure. Packages should have adequate ventilation so that produce will not warm up because of heat arising from respiration. Excessive ventilation, however, may result in wilting.

I. **Pre-cooling**: is an essential step prior to storage at low temperatures. It is the rapid removal of field heat from the produce to reduce the rate of respiration, microbial activity and refrigeration load. Pre-cooling can be done with chilled water, ice or cool air (forced air cooling), whereby the produce is cooled to the half cooling temperature.

J. **Storage:** at low temperatures has been an effective mean of extending the shelf-life of fresh fruits and vegetables. It also enables orderly marketing and distribution of produce in time of peak production. Temperature requirements for different produce may vary depending on variety, location, stage of maturity and other factors. It is important to note that cool storage is a tool used to maintain quality but not to improve it.

K. **Transportation:** Proper handling of the produce during transportation is essential to reduce losses to a minimum and to maintain their quality from the farm to the packing house and from the packing house to market. The used of refrigerated trucks to transport highly perishable and high-value produce will maintain their quality over an extended duration. In non-ventilated vans, the temperature of the fruits or vegetables rises quickly, increasing respiration and decay.

Pack houses serve as a collection center for fruits and vegetables prior to distribution and marketing. The houses can be simple packing sheds with a limited equipment and minimal operations or a large complex that is well equipped and with facilities for specialized operations.

Vacuum Packaging

Sealing of product in an impermeable membrane after the process of removal of air inside the membrane is the process of vacuum packaging and this process is considered as modified atmospheric packaging which is widely used and accepted technique. This method is mainly used to increase the stability of food product more than that of normal refrigeration life and also prevent the browning reaction occurs on the surface of the high perishable products during cutting or damaging by several actions [3]. Oxygen is the major requirement for bacteria, yeast and mould to grow. By removing the air that surrounds food will inhibit the growth of bacteria. The oxygen level drops continuously when the air is removed this will lead to the level of carbon dioxide gets increased and this environment is not suitable for the growth of spoilage microorganisms it will be good economical method to improve shelf life of various food product. Mostly the perishable food that prone to transport and storage are packed by this method and this packaging sometimes need refrigeration because some organisms are resistant to carbon dioxide. This process has several advantages, it reduces the product shrinkage and the moisture loss, so the mass of the product will not affect until the package gets damaged. Vacuum packaging pouches are made of finest virgin material in a thickness suitable for handling, shipping, and storing with minimal care (**Figure 5**).

Figure 5: Vacuum Packaging of Foods Gas Flushing

Gas Flushing

Modified atmospheric packaging is being used with different inert gases, mainly with nitrogen and carbon dioxide. This process is known as gas flushing. These gases act as an inhibitor for several spoilage microorganism in several ways of elevating oxygen from the atmosphere of product **(Figure 6)**.

Figure 6: Packaging of Red cow peas by using gas flushing.

Carbon Dioxide Atmospheric Packaging

Carbon dioxide is already been recognized as effect of inhibiting pathogen, when this gas is used in certain concentrations. This gas must be used in the range of 30-40 per cent if this level exceeds, it will lead to discolouration of several products such as meat. Carbon dioxide sometimes absorbs by the product will results in pack collapse and this affect the moist rate. To overcome this problem, the combination of carbon monoxide within the limit of 35ppm is used. Myoglobin is a compound produce carboxymyoglobin when it combines with carbon monoxide and produce red pigment gives the color of bright cherry. So, the discolouration of food gets reduced by using this method. Several clostridium pathogens will not affect by this method, it will only inhibit the growth of some spoilage microorganism.

Nitrogen Packaging

Nitrogen displays low or no antimicrobial activity on its own. This composition prevents pack collapse because this composition can't have very low capability to dissolve in water and fat when it is subjected to modified atmospheric packaging. Nitrogen is added by displays oxygen in required level during packaging, it delays the oxidative rancidity and growth of aerobic microorganism. It is mainly used in nuts and perishable food packaging. It also acts as a filler gas and keeps flexibility of packaging **(Figure 7)**. So far there is no certain limit for usage of this gas in several foods packaging because it does not produce any adverse effect on food. For fresh cut vegetables the packaging uses 100 per cent nitrogen and after 5 days of storage at low temperature it shows that that the level of oxygen and carbon dioxide as 1.2-5 per cent and 0.5-3.5 per cent a very low amount of adverse effect.

Figure 7: Cumin packaging by nitrogen as a filler.

Argon Packaging

In several laboratories the use of argon for modified atmospheric packaging has been tested. From these results as a reference, in Europe the argon packaging is tried for several different food products and then it is prone to commercial uses. There are number of reasons for using argon, it is much denser than that of nitrogen (1.43 times). The rate of transfer of oxygen is in high rate than nitrogen. During packaging with nitrogen, it is found that the trace amount of oxygen remains, it causes food to oxidize. To overcome this problem argon is used and it fills space completely **(Figure 8)**. This gas also used to prevent the flavour deterioration by inhibit microbial oxidase enzyme that is responsible for oxidation.

Figure 8: Argon packaging of sweet pan.

Controlled Atmospheric Packaging

It is the process involved in maintaining the atmospheric composition. This controlled atmospheric packaging involves in maintaining several parameters mainly temperature and relative humidity and also in control its optimum range, which is major factor involves in instability of food in case of its quality and safety. Several different techniques are evolved to maintain the stability of packaging and food which involves coating using several films, packaging using gases etc.

Modified Atmosphere (MA)

A modified atmosphere (MA) implies removal of, or the addition of gases, resulting in an atmospheric composition different from the one normally existing in air. For example, the N_2 and CO_2 levels may be higher, and the O_2 levels lower than those found in a normal gaseous atmosphere (78% N_2, 21% O_2, and 0.03% CO_2). In this type of storage, the CO_2 and O_2 levels are not controlled under specified conditions.

Appropriate use of MA can supplement refrigerated storage for some products, which could translate into considerable reduction in post-harvest loss. Major benefits can be obtained from its use:

- Reduction in senescence associated with biochemical changes, such as reduction in respiration rate and ethylene production, softening, and compositional changes in fresh produce.
- Decreased sensitivity of fruit to ethylene action at levels of O2 and CO2 below 8 per cent and 1 per cent, respectively.
- Can relieve some physiological disorders such as cooling damage in a variety of products.
- Can have a direct or indirect effect on post-harvest pathogens and insect control.

Some Disadvantages of MA Include

- Initiation of physiological damage, such as black spots in potatoes.
- Irregular maturation of certain fruits, such as bananas and tomatoes (O2 < 2% and CO2 > 5%).
- Abnormal development of flavors and odors at low O2 concentrations (anaerobic conditions).
- Increase susceptibility to diseases.
- Stimulation of sprouting and delay of epidermis development in roots and tubers (e.g., potatoes).

Edible Films

To maintain the nature of product the coating on the outer layer of food material is followed and the materials used for enrobing is edible films. Nowadays food items are not consumed immediately after harvesting, it undergoes several different processes within they reached to consumers such as transportation, storage, handling of product, different stresses are applied on the product. To prevent the loss of product composition, stability and nutritional loss that starts immediately after harvesting these edible coating is used. This will slow down the deterioration **(Figure 9)**.

Figure 9: Efficiency of edible coating.

Polysaccharide Based Coating and Films

The process of coating involves several different polysaccharides, the formulations used are starch, chitosan, cellulose, pectin, and alginate. The strength and structural integrity were provided by the films which acts as a barrier for oxygen aroma and oils but he drawback is it's a hydrophobic nature which likes to hold moisture. Because of their hydrogen bonded network structure and low solubility, tightly packed it act as an oxygen barrier.

Protein Based Coating and Films

Fruits and Vegetables are easily undergone oxidation, to prevent its aroma nature and from contact of oxygen proteins are used for coating such as egg albumen, whey protein, casein, corn-zein, etc., but these are not acts as a moisture barrier. The presence of side residues of amino acid inhibits polyphenol oxidase. Their owing to its network structure by hydrogen bonds, very low soluble nature it is used as a oxygen barrier. The presence of fatty acids in several proteins acts as a moisture barrier, the best example is a Whey protein. Compared to polymeric material degradation of protein is quite simple process and these are from renewable sources. Proteins obtained from milk prevents browning reaction in fruits and vegetables.

Lipid Based Coating and Films

There are several different types of lipids- based material such as paraffin wax, selective types of oils, surfactants etc., are used for coating and these are acts as a effective barrier protect foods from agents that cause deterioration. This type of coating is mainly used to prevent the entry of moisture and water vapour because of its low polarity. The microscopic pores and higher solubility and diffusivity left a characteristics of partial oxygen barrier. Even though it has various advantages these are inflexible and opaque.

Composite Coating

This type of coating can be easily understood by the word 'composite', these types are composed of two or more types of coating materials. There is effective only when they are in good composition. Several proteins exhibit good mechanical properties, from this we can develop stable film system such as plasticized protein. From the research it is observed that by protein casein from milk and acetylated monoglyceride from lipid for apple and potatoes that are slightly processed is stable for up-to 3days from moisture loss and oxidative browning.

Waxing: During washing, fresh vegetables (also fruits) lose part of their outer layer of wax, which protects against humidity loss. As a result, waxing is re-established after washing with an artificial layer of wax that has adequate thickness and consistency to improve appearance and to reduce the loss of water.

Classification: The main objective of this operation is to attain a uniform product for the market. Fresh vegetables are classified by size, weight, or degree of maturity. Classification by size can be done manually in small packing houses with trained personnel. In mechanized packing houses, operations are carried out with perforated belts, divergent belts or cylinders, and sieving. Sorting by mass is usually done electronically but some manually operated machines can classify different weights by a tipping mechanism. Classification by degree of maturity can be done using color charts or by optical methods.

Labelling: Commercial fresh vegetables as well as fruits can be labelled individually with automatic adhesive stickers to identify the product brand, farmer, or retailer. This is extremely important when exporting produce to other countries.

3

Peeling, Slicing, Cubing, Cutting and Other Size Reduction Operations for Fruits and Vegetables

1. **Peeling:** It is one of the integral parts of a food processing, and most agri-cultural crops need to be peeled to remove at the initial stage of food processing Peeling removes inedible portion (peel, seeds, and stalk) of fruits and vegetables. However, the susceptibility to spoilage increases due to acceleration of physiological process and the exposure of the tissues to microorganisms. The shelf-life and quality of fresh-cut produces could be compromised with the peeling.

   The goals of optimum peeling operation are i. Minimizing product losses, ii. Types of products e.g., potato products, iii. Minimizing heat ring formation e.g., apple, potato, iv. Minimizing energy and chemical usage, v. Minimizing the environmental pollution.

   Peeling operation can be grouped under following categories:

   Manual peeling (knife or blade), Mechanical peeling (abrasive devices, devices with drums, rollers, knives or blades and milling cutters), Chemical peeling, Enzymatic peeling, and Thermal peeling (Flame or dry heat peeling, steam or wet heat peeling, thermal blast peeling, and vapour explosion or vacuum peeling.

2. **Cutting:** Cutting, slicing, dicing, and shredding are non-thermal food operation for size reduction. This process reduces the preparation time by consumers. Cutting removes inedible and discoloured portions from foods using knife, chopper, and slicer. However, in this food processing injured tissues are removed so that these are unavailable for microbial spread. The cut tissues result in reduced respiration and enzymes activity, thus retarding rapid spoilage and increases shelf life. The cutting process accumulates fluids on the cut surface, increases microbial load and enzymes activity.

3. **Size reduction operations:** Size reduction is a unit operation involving such activities as cutting, slicing, milling, or pulping of food. It requires energy input to overcome a breaking stress. When this happens, the food will break along lines of weakness resulting in release of energy and sound. So, think of the last time you ate an apple, that sound that you heard is energy being given off. Where enough energy is not added to the material for breaking, it returns to its original size. This is a characteristic of elasticity. Therefore, to break materials, the energy input must overcome both the elastic stress limit and breaking stress.

   Vegetables are subjected to several preliminary operations before processing and after harvesting. As a result of peeling, grating, and shredding, produce will change from a relatively stable product with a shelf life of several weeks or months to a perishable one with a shelf life as short as 1-3 days at chill temperatures. The major preliminary operations include:

   Washing: Root vegetables are washed first to remove all field dirt and to allow inspection.

   Inspection: Vegetables are inspected for quality to comply with consumer demands.

   Selection: Vegetables are selected and graded on a basis of firmness, cleanness, size, weight, colour, shape, maturity, mechanical damage, foreign matter, disease, and insects. This operation can be done manually, or by employing a variety of separation machines to separate and discard unfit produce.

Subsequent Operations

Peeling, cutting, and shredding: Some vegetables such as potatoes and carrots require peeling. Ideal peeling is done very gently, by hand with a sharp knife. It has been reported that hand peeling of carrots increases the respiration rate over that of unpeeled carrots, by approximately 15 per cent, whereas abrasion peeling almost doubles the respiration rate compared to hand peeled carrots. Carborundum abrasion peeled potatoes must be treated with a browning inhibitor, whereas washing is enough for hand peeled pot. Guidelines for pre-peeled and sliced potatoes are followed:

Raw material : A suitable variety or raw material lot should be selected using a rapid storage test of a prepared sample at room temperature. Attention must be focused on browning.

Pretreatment:	Careful washing with good quality water before peeling is required. Damaged and contaminated parts, as well as spoiled potatoes, must be removed.
Peeling:	1) One-stage peeling: knife machine. 2) Two-stage peeling: slight carborundum peeling first, followed by knife peeling.
Washing	Washing is done immediately after peeling. The temperature and amount of washing water should be 4-5°C and 3 L/kg potato, respectively. Washing time: 1 min. Observation: microbiological quality of washing water must be excellent. In washing water, for sliced potatoes, it is preferable to use citric acid with ascorbic acid (maximum concentration of both, 0.5%), combined with calcium chloride, sodium benzoate, or 4-hexyl resorcinol to prevent browning.
Slicing:	Slicing should be done immediately after washing using a sharp knife.
Straining off	Loose water should be strained off through a colander.
Packaging	is done immediately after washing in vacuum or in gas mixture (20% CO_2 + 80% N_2). The head space volume of a package is 2 L/kg of potatoes. Suitable oxygen permeability of packaging materials: 70 cm^3/cm^2, 24 hr, 101.3 kPa, 23°C, 0 per cent RH (80 mm nylon-polyethylene).
Storage	Preferably in dark at 4-5°C.
Other remarks	Good manufacturing practices (GMP) must be followed (hygiene, low temperatures, and disinfection).
Shelf-life	Shelf life of pre-peeled potatoes is 7-8 days at 5°C. Due to browning, sliced potato has very poor stability; the shelf life is only 3-4 days at 5°C.

Second washing and drying: Commonly, a second washing is needed after peeling and/or cutting. For instance, Chinese cabbage and white cabbage should be washed after shredding, whereas carrots should be washed before grating. Washing after peeling and cutting removes microbes and tissue fluid, thus reducing microbial growth and enzymatic oxidation during subsequent storage. Washing fruits and vegetables in flowing or carbonated water is preferable than dipping the product into a tank of water. The microbiological and sensory quality of the washing water must be good and its temperature low, preferably < 5°C. The recommended quantity of water used is 5-10 L/kg for produce before peeling/cutting and 3 L/kg after peeling/cutting.

Preservatives can be used in the washing water to reduce microbial load and to retard enzymatic activity, thus improving the shelf life and sensory quality of produce. The recommended dosage for chemical preservatives in washing water is 100-200 mg/L of chlorine or citric acid. These levels are effective in the washing water before, after, or during cutting to extend the shelf life. However, when chlorine is used, vegetable materials require a subsequent rinse to reduce the chlorine concentration to the level of drinking water and to improve the sensory shelf life. The effectiveness of chlorine should be improved by using low pH, high temperature, pure water, and correct contact time.

Some proposed alternatives are chlorine dioxide, ozone (O_3), trisodium phosphate, and hydrogen peroxide. The use of hydrogen peroxide (H_2O_2) as an alternative to chlorine for disinfecting freshly cut fruits and vegetables shows some promise. H_2O_2 vapour treatment appears to reduce the microbial population on freshly cut vegetables such as cucumbers, green bell peppers, and zucchini. Moreover, H_2O_2 vapor treatment extends the shelf life of vegetable products without leaving significant residues or causing loss of quality. However, more research is needed to optimize H_2O_2 treatments about efficacy in delaying the growth of spoilage bacteria in a wide variety of vegetable products.

Module-3

1

Minimal Processing

Minimal processing of fruits and vegetables

The technological development and modernization during recent years have resulted in in organisation coupled with Lifestyle modification and changes in dietary habits. This development and Organisation have made it easier to meet the requirement of processed food. Processing of fruits and vegetables effects shelf life providing ready to eat food with similarities to fresh product. Before the food Industries shoulder a bigger responsibility of utilising a novel technology to produce safe food without detrimental effects on quality. The industries need to who built biological (enzymatic / fomentation and bio-preservation) and Engineering solutions like non-thermal high-pressure treatment, cook chill, microencapsulation, vacuum impregnation etc. For the adoption of de - constructive culinary techniques such as 3D food printing and Molecular gastronomy as well as developments in nutrigenomics and digital technologies may facilitate novel product format, personalization, and access to niche market. Concretely, e - operating movement from Naresh mint to health improvement in present conditions demand or shift from defensive market oriented offensive market developing strategies including collaborative networks with Research Organization.

It is also defined as any fruit and vegetable that has been subjected to different processing steps (e.g. peeling, trimming, cutting, washing, disinfection rinsing etc.).

Minimally processed products are defined as food product that undergo very little processing keeping in text its colour, flavour, appearance, texture and taste. It is an optimized design for preserving food in the aspects of providing nutritional safety and sensory quality without the use of synthetic additives. Minimal processing concept emerged out due to increasing demand of consumers for natural and fresh like foods with desired shelf life. In food processing the term technology is also used to express the 'technique' which allows minimal physicochemical, oxidative, and mechanical damage to the food products. Hence the aims of minimal processing include:

- To make the food safe chemically and microbiologically
- To retain the desired flavour, colour and texture of food product
- To provide convenient to the consumers and enhanced shelf life
- To use low severity of processing methods
- To maintain quality as fresh or close to the fresh prepared meals or products.

Applications of Minimal Processing

Various techniques or combination methods are used in to the undergo minimal processing. To improve the organoleptic characteristics and to maintain nutritional value the traditional preservation methods are optimized to obtain a fully edible product while providing convenience and functionality of the food product.

Sometimes mild heat or no heat methods are used for microbiological treatment. These techniques are categorized as novel preservation techniques. Lastly synergistic effect that is combination of various methods is used.

Broadly minimal processing can be divided into two categories: First are based upon plant source such as fruits and vegetables while the other is animal based such as meat, fish and seafood. Besides some other category is like part bread, ready meal, cook chill, cook freeze also are appearing.

Plant Based Minimally Processed Foods

High moisture content enzymatic activity in fruits and vegetables make them prone to microbiological damage. This calls for the concern to follow standard procedures and equipment for grading, sorting, washing, feeling, shopping, and shedding. The presence of enzymes like polyphenol oxidase, galactoseronase, lipoxygenase, etc play an essential role in initiating oxidation process + spoilage of cut surface for diced and sliced vegetables and fruits. In Fast Moving Consumer Good (FMCG) with PH> 6.0, there is possibility of growth of microbes when cut surfaces are exposed to air. But the minimally processed and we are not allowed to undergo treatment abuse and are needed to be stored at chilling temperature of 4-7^{0}C

Animal based minimally processed foods

animal and sea foods are more prone to infection by microorganisms as compared to plant food. Combination methods can create vendor for in activating the microbes and also to maintain nutritive value. Also, typical spoilage of meat products and seafood may occur by denaturation of protein, which further involve the dissociation of their structures, protein aggregation and gelation. A combination of methods including heat and non-thermal techniques along with refrigeration storage can result in a convincing strategy.

Techniques of Minimal Processing

Minimally processed operations include all those procedures such as washing, sorting, trimming, peeling, slicing, chopping, antioxidant treatment, packing etc. Which do not hamper the fresh like quality of fruits and vegetables. These fruits and vegetables require no additional preparation for their use the results are that these products can be consumed straightway in very short time. The most used terms used for minimally processed products include lightly processed, partially processed, ready to eat or ready to cook. These organisms including different species of bacteria and microbial growth on minimally processed product can be controlled by good sanitation, achievement the Hazard Analysis Critical Control Point (HACCP) proper temperature management during and after shower stainless steel knives, use of proper packing material, the low Oxygen and elevated carbon dioxide level sometimes in conjunction with carbon monoxide or nitrogen retards the microbial growth. The use of low temperature storage (0 to 50^{0}C), chemical preservatives such as chlorine (140 ppm), ascorbic acid at 3-5 per cent and sodium metabisulfite at 200 ppm, reduction of water activity by centrifugation and modified and controlled atmosphere helps in retaining the quality and extend the storage Life by inhibiting metabolic activity, decay and ethylene production.

Effect of Minimal Processing on the Quality of Food Products

The appearance, taste, colour and texture of fresh cut fruits and vegetables are the most appealing attributes. Refresh that product are most preferred in the today's scenario due to the convenience they offer to the consumer, their availability at different portion sizes, employment of reduced labour in preparation and consumption and greater retention of nutrients and Aroma. The cutting and rounding of minimally processed products render them vulnerable to microbial infection and enzymatic Browning with higher metabolic activities. But these injuries fast and the respiration rate which for the triggers the increased production of Ethylene and finally senescence. Hence, several methods employed in minimal processing to prevent in natural characters of fruits and vegetables are as follows:

a) **Vacuum packaging**: It is the sealing of the product in an impermeable membrane after removal of air. stability of the product is increased once Browning is prevented.

b) **Gas Flushing**: Atmosphere is modified by using different inert gases in particular nitrogen, carbon dioxide and Argon. These gases inhibit the activity of several microorganisms by elevating oxygen from the atmosphere of the product.

c) **Edible coating:** Outer layer of the food product is enrobed with edible films of polysaccharide, protein, lipid or of composite coating. This help in preventing the entry of moisture and water vapour because of its low polarity.

Recently, the concept of minimal processing of food has assented to higher levels because of consumers. In response to consumers increasingly demand of natural and fresh like foods minimal processing Technologies like Ozone treatment, chlorine treatment etc are used to delay the deterioration of the minimally processed food products.

Quality Attributes for Minimally Processed Food

Quality of a product serves an important point in the market of minimal processing foods, the most noted quality attributes are safety of product, colour, flavour, nutritional value, appearance. These are based on the external and the internal factors but it is determined by the conditions of pre and post harvesting. Variety of plant belongs, ripening and maturity stage, these are taken into consideration before processing of foods because these play a vital role in marketing of foods.

Appearance

Appearance is considered as the most important quality attributes that exposes freshness and also attraction, these attributes falls into color dispersion, glossiness or exposure of product, defects in shape or structure of product. Colour change happens mainly due to ripening and also discoloration is due to oxidation, enzymatic browning when fresh fruits and vegetables are subjected to cutting. The other aspects are skin glossiness, wrinkling, blemishes, consistent size and dimension, wilting caused by several pathogens or by other environmental causes.

Texture

The texture of food products is often showed by using specific terms such as firmness, toughness, crispness and sometimes juiciness. The nature of product can be easily collapsed by deterioration of products by several factors. This stability of structure can be easily identified by sense of touch, chewing of product and mouth feel. If it is oxidised the texture get collapsed and exhibit unusual low rigidity to the foods it may lead to intolerant smell. The crispness of product can also be caused by dehydration but it should not be taken into an account. Sometimes the quality of product affect adversely because of loss in juiciness of product.

Flavour

Flavour is due to the presence of free volatile components present in the sample. The most common tastes observed are bitter, sweet, astringent and acid taste. Ripening of fruit often defined by the sweetness of fruit. Acid level in the food also very useful in impart flavour and it is often affected during ripening and post-harvest stages. Bitterness is the off-flavour produced during improper storage conditions. Coldness also a major factor that limits or acts as a barrier for aroma development.

Nutritional Content

The most important source of dietary fibre, minerals and vitamins are fruits and vegetables. The nutrition content of fruits and vegetables gradually decrease immediately after harvesting. To prevent these losses several techniques such as storage and packaging of materials should be considered. The loss of vitamins and minerals and some other compounds are due to cutting because it leads to the secretion of ethylene and this will be the main cause for increase in respiration and senescence. The most degradable vitamin is ascorbic acid, it is considered as the index for freshness this is mostly unstable in several fruits and vegetables.

Nowadays several different techniques are used to produce minimal process foods. The use of non-thermal techniques such as pulsed electric field, ultraviolet light, Ultrasonication, high pressure processing produces some adverse effect on the nature of product because of the mild production of heat and pressure applied during processing. Other techniques such as UV-C light treatment (200- 280 nm) disinfectant the product but it will affect the flavour and surface rigidity and water-jet method is the industrial method for cutting of fruits and vegetables also increase the moisture content by absorption of water through pores of fruits and vegetables. From this it is concluded that these techniques explained in this section, do not affect the composition and nature of the food product when compared to all other techniques and can be recommended for use in small scale and large-scale industries.

2

Fruit Preservation Through Thermal Processing

There is wastage and loss of food during its journey from production to consumption. Approximately one-third of the massive 1.3 billion tonnes of food produced annually goes to waste due to various factors such as poor handling after harvesting, unfavorable storage conditions, insect infestations, and diseases. To combat these losses, synthetic chemicals have historically been used to control pathogens and pests, but concerns over their potential health risks have led to restrictions on their usage. In response, researchers are now focusing on developing safer and more effective physical treatments for disinfecting and protecting horticultural produce. This chapter specifically explores the advantages of using heat treatment as one such method, discussing how it helps maintain the quality of fruits and vegetables while simultaneously reducing postharvest losses.

Some of the heat processing methods are discussed below:

Fruit juices, RTS and nectars are preserved by pasteurization or by using chemical preservatives. Squashes, crushes, syrups and cordials are preserved by adding chemical preservative like potassium metabisulphite or sodium benzoate.

1. **Pasteurization:** Preservation of fruit juices by application of heat is the most common method. Pasteurization is a process in which juice is heated to 100^{0}C or slightly below for a sufficient time to inactivate/kill the micro-organisms, which cause spoilage. Usually, the fruit juices are pasteurized between 75 and 88^{0}C with times ranging from 30 sec to 30 min depending on the type of heating system, the nature of the juice and the size of the container. Pasteurization can be performed either by heating at low temperature for a long time (LTLT) or heating at high temperature for short time (HTST). Commonly followed methods are:

   a) **Holding pasteurization:** In this method, the prepared juice is filled into glass bottles with a proper head space and the bottles are sealed airtight. After sealing the bottles are pasteurized. However, this method is not followed for commercial scale products.

b) **Pasteurization by overflow method:** In this method, the juice is heated to a temperature about 2.5^{0}C higher than the pasteurization temperature and filled into the hot sterilized bottles up to the brim. Care should be taken to maintain the temperature during filling and sealing operation. The sealed bottles are processed in boiling water for specified period. After processing the bottles are cooled. On cooling, the juice contracts to leave a small headspace in the bottle which does not contain any air. The method is commercially followed in preservation of all types of juices in the food industry.

c) **Flash Pasteurization:** In this process, the juice is heated only for short time at a temperature about -5.5^{o}C than the pasteurization temperature, filled into the containers and sealed air tight under cover of a steam to sterilize the seal and then cooled. This process is also known as HTST (High temperature and short time) method and the juice is heated vary rapidly. Flash pasteurization helps to minimize the loss of flavour, better retention of vitamins, keep juice uniformly cloudy and minimize the cooked taste of the juice.

2. **Aseptic processing and packaging of fruit juices:** Aseptic processing and packaging is defined as the process in which a commercially sterile product is packed into pre-sterilized container in a sterile environment. The system make use of high temperature short time (HTST) sterilization in the temperature range of 90- 110^{0}C for acid products ($pH<4.6$) and ultra-high temperature (UHT) sterilization 121^{0}C and above for low acid foods ($pH>4.6$). The commercial aseptic sterilization process takes place in a continuous, closed system. Aseptic processing may produce products with better retention of nutrients and excellent sensory quality. Apple, mango, litchi, pineapple drinks etc. in tetra pack are processed commercially using aseptic processing and packaging system.

3. **Preservation with chemical**: Fruit juices, pulps, squash, cordial, syrup, Ready-To-Serve (RTS) drinks etc., are preserved with chemical preservatives. Fruit juice and pulps in bulk are preserved with chemical preservatives. Two chemical preservatives most commonly used in preservation of fruit and vegetable products are:

i) **Benzoic acid**: Benzoic acid is the effective agent but sparingly soluble in water, thus its sodium salt, which is water soluble, is generally used. Benzoic acid is more effective against yeast as compared to moulds. However, it does not stop lactic acid and acetic acid fermentation. The quantity of sodium benzoate required depends on the nature of the juice, its acidity and type of microbial infection. The limit of benzoic acid to be added in various products as per FPO (Fruit Products Order) is given in Table 1.

ii) **Sulphur dioxide**: Potassium meta-bi-sulphite ($K_2O_2SO_5$ or $K_2S_2O_5$) is commonly used as a source of sulphur dioxide. On addition to fruit juice or beverage it reacts with acid of the juice and form potassium salt and sulphur dioxide, which is liberated and form sulphurous acid with the water of the juice. Sulphur dioxide ismore effective against mould spores and bacteria than yeast and also inhibits enzymes etc. It acts as antioxidant and bleaching agent thus help in the retention of ascorbic acid, carotene and other oxidisable compounds. It also retards non-enzymatic browning or discoloration of the product. Its effectiveness depends on the acidity, pH, temperature and other substances present in the fruit juice.

The SO_2 should not be used in naturally colored juices like phalsa, Jamun, pomegranate and strawberry juices, on account of its bleaching action. It should also not be used in those juices, which are to be packed into tin containers, because it may act on the tin of the containers causing pinholes, forms hydrogen sulphide and black compounds.

The potassium meta-bi-sulphite should be first dissolved in a small quantity of water and then added to the juice to be preserved. The limit of SO_2 to be added in different processed products according to Food Safety and Standards Regulation 2011 and FPO specification is given in Table 1.

Table 3: Limits for permitted preservatives in fruit juice and beverages as per Fruit Products Order

Fruit juice/ beverages	**Preservatives**	**Part per million(ppm)**
Fruit pulp or juice for conversions into jams and other products		
Cherries	SO_2	3,000
Strawberries and raspberries	SO_2	2,000
Other fruits	SO_2	1,000
Fruit juice concentrate	SO_2	1,500
Squashes, crushes, fruit syrup, sharbats, cordials, fruit juice and barley water	SO_2 or Benzoicacid	70
Sweetened ready to servebeverages (RTS)	SO_2 or Benzoicacid	120

4. **Preservation by sugar:** Fruit juice containing 66% sugar generally does not ferment. Fruit syrup or sharbats with high total solids (65% and above) have a very low water activity hence micro-organism do not grow. The sugar acts as a preservative by osmosis and does not support the growth of micro-organism. However, the growth of mould and yeast can occur on the surface of jams or jellies which need to be protected by using airtight packing or covering the product with molten paraffin wax.
5. **Preservation by freezing:** The properly frozen juice retains its freshness, colour, taste and aroma for a long time. De-aerated juice is transferred into containers, which are hermetically sealed and frozen. Freezing does not sterilize fruit juices; it merely reduces the temperature to the point where micro-organisms do not multiply and chemical changes take place very slowly. The more rapidly a juice is frozen, the smaller the ice crystal formed and the less the amount of colloidal matter coagulates. When the juice is frozen to - 18^{0}C, practically all of the juice will be solid except for a little thick syrup in the center of the containers. This method is particularly useful in the case of juices whose flavour is adversely affected by heating.
6. **Preservation by drying**: Micro-organisms need water for their growth and multiplication and as the water is removed, they are unable to grow. Fruit juices can be preserved in the form of powder by different methods viz., spray drying, drum drying, freeze- drying, foam-mat drying etc. Fruit juice powders are highly hygroscopic and require special care in packing. It should be packed in free-flowing powder form in hermetically sealed containers with in package desiccant to prolong the shelf-life of the product. Lemon juice powder can also be prepared by using foam mat drying process.
7. **Preservation by carbonation**: Carbonation is a process of mixing carbon dioxide under pressure with water, juice or beverages so that the product when served; gives off gas in fine bubbles and has the characteristics taste. Carbonated beverages are generally bottled with carbon dioxide content ranging from 1 to 8 g/L. Though this concentration is much lower than that required (14.6 g/L) for complete inhibition of microbial growth, yet it is significant in supplementing the lethal effect of acidity on pathogenic bacteria. Carbonation also helps in removal of air which creates anaerobic condition and reduces the oxidation of ascorbic acid. Since moulds and yeasts require oxygen for their growth and as such become inactive in the presence of carbon dioxide. Thus, the absence of air and presence of carbon dioxide in fruit beverages help to prevent the growth of moulds and yeast. The keeping quality of carbonated

beverages is enhanced by adding about 50ppm sodium benzoate. Carbon dioxide in beverages is measured in terms of gas volume. The gas volume is defined as the amount of gas in milliliters that a given volume of water will absorb at atmospheric pressure i.e. 760 mm of Hg at 15.5^{0}C temperature. Generally, 1.3-4.0 gas volume CO_2 is used for carbonation of fruit juice. Lime, lemon, grape, apple juice can be carbonated by using carbonation process.

8. **Preservation by filtration:** In this method, the clarified fruit juices (particularly apple and grapes) are passed through special filters, which are capable of retaining yeasts and bacteria. Various types of germ proof filters are used for this purpose and the method is used for soft drinks, fruit juices and wines.

Filling and processing: Bottles are thoroughly washed with hot water and filled leaving 1.5-2.5 cm headspace. The bottles meant for heat processing are sealed by using crown corks whole those preserved by using chemical preservative can be sealed by using PP (pilfer proof) caps.

Blanching Operations and Equipment

Blanching is used to destroy enzymatic activi ty in vegetable and some fruits prior to other processing like freezing or dehydration or canning or thermal processing. It is a pretreatment by mild heat for a specific time followed by rapid cooling or passing immediately to the next processing stage. The time and temperature combination varies from product to product, the condition and size of product. Generally, the temperature varies from 88 to 99^{0}C. In some of the fruits and vegetables poly phenol oxidase enzyme is responsible for discoloration in presence of oxygen, hence it needs to be inactivated by blanching pretreatment, before further processing of fruits and vegetables to maintain its original color after processing.

Mechanism of Blanching

Plant cells are discrete membrane-bound structures contained within semirigid cell walls. The outer or cytoplasmic membrane acts as a skin, maintaining turgor pressure within the cell. Loss of turgor pressure leads to softening of the tissue. Within the cell are a number of organelles, including the nucleus, vacuole, chloroplasts, chromoplasts and mitochondria. This compartmentalization is essential to the various biochemical and physical functions. Blanching causes cell death and physical and metabolic chaos within the cells. The heating effect leads to enzyme destruction as well as damage to the cytoplasmic and other membranes, which become permeable to water and solutes. An immediate effect is the loss of turgor pressure. Water and solutes may pass into and out of the cells, a major consequence being nutrient loss from the tissue. Also, cell constituents, who had previously been compartmentalized in sub cellular organelles, become free to move and interact within the cell.

The following factors affect the blanching time

1. The type of fruit or vegetable.
2. The size of the pieces of food.
3. The blanching temperature and
4. The method of heating

Purpose and Objective of Blanching

The purpose of blanching is to achieve several objectives.

1. To soften the tissue to facilitate packaging.
2. To avoid damage to the product.
3. To eliminate air from the product.
4. To preserve the natural colour.
5. To destroy or retard certain undesirable enzymes.
6. To help preserve natural flavour.

The major purpose of blanching is frequently to inactivate enzymes, which would otherwise lead to quality reduction in the processed product. For example, with frozen foods, deterioration could take place during any delay prior to processing, during freezing, during frozen storage or during subsequent thawing. Similar considerations apply to the processing, storage and rehydration of dehydrated foods. Enzyme inactivation prior to heat sterilization is less important as the severe processing will destroy any enzyme activity, but there may be appreciable time before the food is heated to sufficient temperature, so quality may be better maintained if enzymes are destroyed prior to heat sterilization processes such as canning. It is important to inactivate quality-changing enzymes that are enzymes which will give rise to loss of colour or texture, production of off odours and flavours or breakdown of nutrients.

Many such enzymes have been studied, including a range of peroxidases, catalases and lipoxygenases. Peroxidase and to a lesser extent catalase are frequently used as indicator enzymes to determine the effectiveness of blanching. Although other enzymes may be more important in terms of their quality-changing effect, peroxidase is chosen because it is extremely easy to measure and it is the most heat resistant of the enzymes in question. More recent work indicates that complete inactivation of peroxidase may not be necessary and retention of a small percentage of the enzyme following blanching of some vegetables may be acceptable. Blanching causes the removal of gases from plant tissues, especially intercellular gas. This is especially useful prior to canning where blanching helps achieve vacuum in the containers, preventing expansion of air during processing and hence reducing strain on the containers and the risk of misshapen cans and/or faulty seams. In addition, removing oxygen is useful in avoiding oxidation of the product and corrosion of the can. Removal of gases, along with the removal of surface dust, has a further effect in brightening the colour of some products, especially green vegetables.

Shrinking and softening of the tissue is a further consequence of blanching. This is of benefit in terms of achieving filled weight into containers, so for example it may be possible to reduce the tin plate requirement in canning. It may also facilitate the filling of containers. It is important to control the time/temperature conditions to avoid overprocessing, leading to excessive loss of texture in some processed products. Calcium chloride addition to blanching water helps to maintain the texture of plant tissue through the formation of calcium pectate complexes. Some weight loss from the tissue is inevitable as both water and solutes are lost from the cells.

A further benefit is that blanching acts as a final cleaning and decontamination process. It also removes pesticide residues or radionuclides from the surface of vegetables, while toxic constituents naturally present (such as nitrites, nitrates and oxalate) are reduced by leaching. Very significant reductions in microorganism content can be achieved, which is useful in frozen or dried foods where surviving organisms can multiply on thawing or rehydration. It is also useful before heat sterilization if large numbers of microorganisms are present before processing.

Principles of Blanching

Blanching is achieved in hot water for a short period of time or in an atmosphere of steam. In water blanching, the product is moved through water usually maintained at a temperature between 88 and 99 ^{0}C. In steam blanching the product is carried on a belt through a steam chamber into which live steam is constantly injected. The steam chamber is hooded and equipped with exhaust and also a drain for the condensate. The time temperatures are regulated for each specific product to achieve the desired enzyme inactivation, colour preservation and other characteristics. As a guide, the operator utilizes either the catalase or the peroxidase tests to determine the adequacy of blanching. Currently the peroxide test is commonly used in industry. For the most part, a negative peroxidase test is necessary to prevent the development of undesirable characteristics in the finished product. Immediately after blanching, vegetables are quickly cooled, usually in cold water, which often serves as means to convey the product to the next operation. A rod type cylindrical reel connected to the discharge of blancher and equipped with water sprays also serves as an excellent cooling system.

Processing Conditions for Blanching

It is essential to control the processing conditions accurately to avoid loss of texture, weight, colour and nutrients. All water- soluble materials, including minerals, sugars, proteins and vitamins, can leach out of the tissue, leading to nutrient loss. In addition, some nutrient loss (especially ascorbic acid) occurs through thermal lability and, to a lesser extent, oxidation. Ascorbic acid is the most commonly measured nutrient with respect to blanching, as it covers all eventualities, being water soluble and hence prone to leaching from cells, thermally labile, as well as being subject to enzymic breakdown by ascorbic acid oxidase during storage. Wide ranges of vitamin C breakdown are observed, depending on the raw material and the method and precise conditions of processing. The aim is to minimize leaching and thermal breakdown while eliminating ascorbic acid oxidase activity, such that vitamin C losses in the product are restricted to a few percent. Generally, steam blanching systems give rise to lower losses of nutrients than immersion systems, presumably because leaching effects are less important. Blanching is an example of unsteady state heat transfer involving convective heat transfer from the blanching medium and conduction within the food piece. Mass transfer of material into and out of the tissue is also important. The precise blanching conditions (time and temperature) must be evaluated for the raw material and usually represent a balance between retaining the quality characteristics of the raw material and avoiding over-processing.

The Following Factors must be Considered for Deciding Processing Conditions of Blanching

1. Fruit or vegetable properties, especially thermal conductivity, which will be determined by type, cultivar, degree of maturity etc
2. Overall blanching effect required for the processed product, which could be expressed in many ways including: achieving a specified central temperature, achieving a specified level of peroxidase inactivation, retaining a specified proportion of vitamin C.
3. Size and shape of food pieces
4. Method of heating and temperature of blanching medium

Time/temperature combinations vary very widely for different foods and different processes and must be determined specifically for any situation. Holding times of 1–15 minutes at 70–100 °C are normal.

Methods of Blanching

The two most widespread commercial methods of blanching involve passing food through an atmosphere of saturated steam or a bath of hot water. Both types of equipment are relatively simple and inexpensive. Microwave blanching is not yet used commercially on a large scale. There have been substantial developments to blanchers in recent years to reduce the energy consumption and also to reduce the loss of soluble components of foods, which reduces the volume and polluting potential of effluents and increases the yield of product. Conventional steam blanching consists of conveying the material through an atmosphere of steam in a tunnel on a mesh belt. Uniformity of heating is often poor where food is unevenly distributed; and the cleaning effect on the food is limited. However, the volumes of waste water are much lower than for water blanching. Fluidized bed designs and 'individual quick blanching' (a three-stage process in which vegetable pieces are heated rapidly in thin layers by steam), held in a deep bed to allow temperature equilibration, (followed by cooling in chilled air) may overcome the problems of non-uniform heating and lead to more efficient systems.

Equipment used for Blanching

a) **Steam Blanchers:** At simplest a steam blancher consists of a mesh conveyor belt that carries food through a steam atmosphere as shown in the figure 10. The residence time of the food is controlled by the speed of the conveyor. In conventional steam blanching, there is often poor uniformity of heating in the multiple layers of food. To overcome this Individual Quick Blanching (IQB) was introduced which involves blanching in two stages. In the first stage food is heated in single layer to a sufficiently high temperature. In the second stage a deep bed of food is held for sufficient time to allow the temperature at the center of each piece to increase to that needed for enzyme inactivation.

b) **Hot Water Blanchers:** There are several different designs of blanchers each of which retain the food in hot water at 70 – 100 ^{0}C for specific time and thus remove it to a dewatering-cooling section. A blancher cooler is shown in the figure 11. It has three sections: a pre-heating stage, a launching stage and a cooling stage. The food is preheated with water that is circulated through a heat exchanger. After blanching a second re-circulation system cools the food. The two systems pass water through the heat exchanger and this heats the pre-heated water and simultaneously cools the cooling water. A re-circulated water- steam mixture is used to blanch the food and final cooling is done by cold air. A counter-current water blancher is shown in the figure 12.

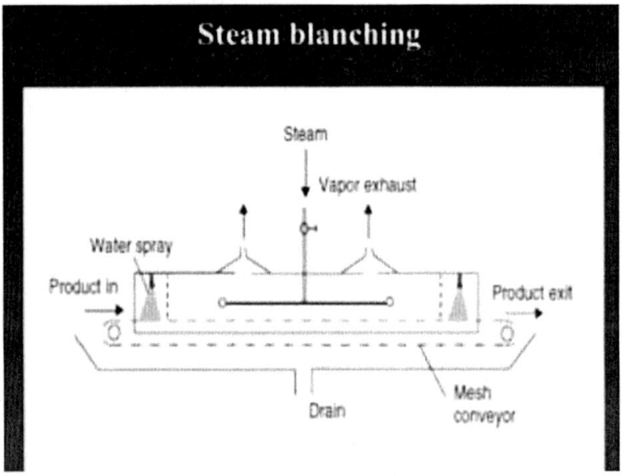

Figure 10: Steam blancher

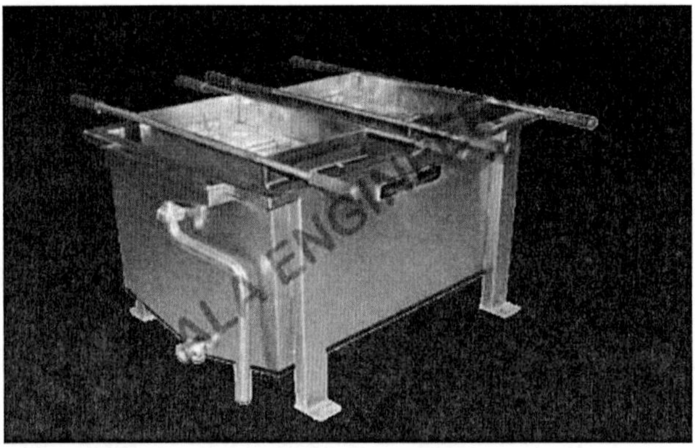

Figure 11: Semi-automatic hot water blancher

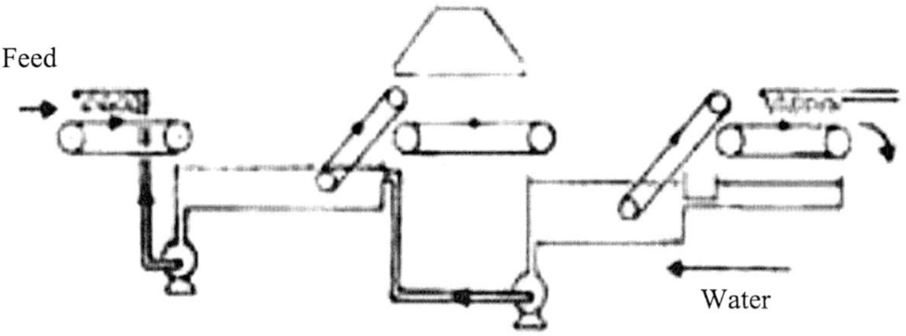

Figure 12: Counter-current hot water blancher

The quality of blanched products depends significantly on the time–temperature combinations of blanching and on the vegetable type. Under-blanching speeds up the activity of enzymes and is worse than no blanching. Over-blanching causes losses in texture, color, phyto-chemicals, and minerals. Typically, industrial blanching processes utilize temperatures ranging from 70 to 95^{0}C and times are usually within 10 minutes whereas for domestic purposes vegetables are generally blanched, or given an extended blanching period which ultimately leads to cooking, for 10–15 min in water at temperatures ranging from 95–100^{0}C.

Generally, blanching is carried out by the application of a wet medium such as steam or hot water to provide uniform heating and a high-heat transfer rate. Both in domestic and industrial processing, several blanching methods may be employed such as conventional water blanching, microwave, or steam blanching; the regime being dictated by the nature of the raw material and the desired properties of the final product. Traditionally, blanching is carried out either at a low temperature (55–75^{0}C) for long-time, typically referred to as LTLT or high-temperature short-time (80– 100^{0}C) for less than 10 min, referred to as HTST depending upon the type of vegetable. Conventional water blanching usually imparts more uniform processing. However, prolonged water blanching results in considerable losses in phytochemicals and antioxidant properties. Microwave heating is three- to five-times faster than conventional heating and relies on the application of dielectric heating. This is accomplished by using microwave radiation to heat water and other polarized molecules within the food, leading to heat generation in the entire volume at the same rate due to internal thermal dissipation of water molecule vibrations in the food. It has advantages over conventional heating methods such as precision timing, speed, and energy saving.

Steam blanching is generally carried out in a steam blancher where the vegetable product is exposed directly to a food-grade steam typically at a temperature close to 100^{0}C. Steam blanching results in minimum losses in phyto-chemicals and antioxidant capacity furthermore, it requires less time than conventional blanching because the heat transfer coefficient of condensing steam is greater than that of hot water and it is proven to be comparatively economical as it saves energy.

Hence it can be said that blanching is an important intermediate thermal processing step to enhance preservation and quality of foods by inactivating enzymes that can cause loss of flavor and color. However, timing of blanching is crucial and should be appropriately adjusted for size and type of foods based on varying susceptibility to degradation with thermal processing. Under blanching stimulates enzymatic actions and increases degradation rate, whereas over-blanching causes loss of texture, color, and flavor qualities. Bioactive phytochemicals can undergo chemical changes during blanching as well. The effect of blanching on chemical changes of bioactive phytochemicals depends on several factors, including the method of blanching, thermal stability of different phytochemicals, enzyme activity, and location of phyto- chemicals in the plant structure.

Canning: Definition, Processing Steps and Equipment, Cans and Containers, Quality Assurance And Defects in Canned Products

Canning process was introduced about two centuries ago, and for long time, it has been one of the main means of food preservation, together with chilling and freezing. Canning of fruits and vegetables started in 1927. The introduction of modern techniques of processing and preservation by addition of chemical preservatives could also be said to have been started from the same time. During the period 1927- 1940, the processing was mainly in the manufacture of soft drinks like squashes, juices, cordial, barley water etc. From 1940 onwards the industry diversified the product mix and it started making canned fruits and vegetables, jams, jellies and marmalade, tomato products, fruit juices etc.

The food canning history began in the late eighteenth century in France when Nicholas Appert discovered that the application of heat to food in a sealed glass container prevented food spoilage. Later, Peter Durand developed a method of sealing food in tin containers; this idea was improved by Bryan Dorkin and John Hall who installed the first commercial cannery in England. Some years later, L. Pasteur gave a reasonable explanation for canning's effectiveness when he demonstrated that microorganisms were responsible of food spoilage. Gradually, the production of canned foods became mechanized and the developments associated to food canning continue today. Conventional canning is a method of food preservation in which a food is placed in hermetically sealed containers and heated to destroy microorganisms. The main objective of heat application is to destroy pathogenic and spoilage microorganisms, and at the same time, the hermetic container prevents contamination by new microorganisms. Although the use of metal containers is common, there are other alternatives as glass jars, plastic cans, and retort pouches. The level of heat applied to a food depends on several factors: acidity of processed food, density, composition, resistance to heat transfer of food, heat resistance of microorganisms of interest in food, initial load of microorganisms, container, heating medium, equipment, processes, etc. After heating, canned foods are cooled and then handled at room temperature maintaining container integrity and preventing recontamination of product.

Concepts Related to Canning

To evaluate the suitable amount of heat for the canning process, several concepts must be studied. Among these concepts, the following are considered: pH is one of the most relevant factors when designing canned foods; the pH of the food plays a key role in determining the extent of heat processing needed to insure a safe and stable final product. The pH value of a food represents the molar concentration of hydrogen ions (in other words, the acidity level of a product); this concentration decreases as the pH value of the food increases. So a low pH value means a higher concentration of hydrogen ions. The range of pH values is between zero and 14. A pH of 7 is neutral (corresponds to pure water), while values less than 7 are acidic and those greater than 7 are basic or alkaline. In the canning industry, a low-acid food is defined as a food having a pH higher than 4.6, while an acid food is defined as a food with a pH value equal or lower than 4.6. This simple classification of foods considers a reference pH value of 4.6. This value is based on *Clostridium botulinum* spores.

Preparation of Foods for Canning

To obtain wholesome canned foods, good manufacturing practices must be used during storage, handling and preparation of raw foods and ingredients. Some problems can appear because of unsuitable handling of foods before thermal process. It is important to minimize the microbial contamination of food in each operation prior to heat treatment; otherwise, the microbial population increases and designed thermal process could be insufficient to reach the condition of thermal sterility. Depending on the characteristics of foods, different preparative actions are taken to process a product. Fruits and vegetables are usually peeled, pitted, de-stemmed and seeded. Some leafy vegetables (like spinach) are heated (blanching) before canning to remove air and facilitate the packing of leaves in containers; in other cases, the heating inactivates enzymes to prevent some changes in product, or it can increase the initial temperature of food. High-acid fruits or vegetables can be thermally processed before they are placed in containers. Usually, meats are precooked, boned, and compacted before thermal processing. In the case of seafood, most fish and shellfish are boned or shelled before packing; however, some smaller fish, like sardines, are packed with their bones, considering that they will soften during the thermal process.

Essentially, there are two categories of thermal processes in the canning industry. One is based on the use of retorts (or autoclaves: conventional canning), and the other is the aseptic processing of foods. In processes using retorts, containers are filled with food, followed by sealing them and then heated using steam under pressure until both container and food become sterilized together. In aseptic processing, a liquid food is sterilized outside the container by means of heat exchangers, and then it is cooled in the same equipment. The cooled sterile food is then filled and sealed in a container that has been previously sterilized; these operations are carried out in a sterile environment at room temperature. Summarizing, the retort processing is an 'in-container' sterilization, which can be applied to all types of foods, while aseptic processing is an 'out-of-container 'sterilization used for liquid foods. Additionally, a third category of thermal processes can be considered; it is the atmospheric processing or pasteurization, applied to acid and acidified foods that require only a mild heat treatment. The thermal process must be specific for each product, container type and size, and type of sterilization equipment. The thermal process is obtained by combining information about thermal destruction of microorganisms and product heating data in the proposed sterilization equipment. The amount of heat needed to destroy an expected number of microorganisms can be determined according to information on thermal resistance of microorganisms of interest in the product and the level of microbial destruction planned with the thermal process.

What is Canning?

Canning is a post-harvest preservation technique, employed by heating jars or containers, filled with food, up to a specific temperature that destroys the micro-organisms and deactivates enzymes, responsible for the food spoilage. During the process of heating air leaves the container and a vacuum seal is formed which prevents the return of air along with contaminating micro- organisms. Fruits and vegetables are primarily canned in the season of abundance and are sold in the off-season resulting in a better return to the producers.

Contrary to pasteurization, canning involves higher temperatures and longer exposure created by steam under pressure, which is the most effective way to kill endospore-forming microorganisms and vegetative cells. The temperature generally exceeds 100^{0}C and the duration for heat exposure mainly depends on the microbial load, acidity of foods, and presence of preservatives. The heat treatment is followed by rapid cooling. *Clostridium botulinum* is the only spore-forming bacteria that survive the canning treatment.

The Canning Process Includes the Following Steps

Preparation of food: includes preliminary operations such as selection of the raw material, washing, grading, sorting, peeling, sub-sizing, and blanching. High-pressure water sprays are employed for washing; sorting and grading are done by visual inspection based on weight, shape, and color; blanching is done by subjecting clean food to hot water or steam to deactivate enzymes, remove respiratory gases, fixing natural color, and pre-heat the product to help in the vacuum formation in the can.

Can Filling

Is the process of aseptically filling whole, sized, or pureed fruit or vegetables in the containers. Before filling, cans are subjected to a steam jet or water to remove any adhered dust particles of foreign material. Tin cans are generally made of thin steel plates of low carbon content, lightly coated on either side with tin metal to a thickness varying from 0.31 mm to 1.54 mm. The different types of metal base plates used for can manufacturing, are as follows:

Type L: It is a high-purity steel with low metalloid and residual content. This kind of base plate is used for highly acidic foods.

Type MR: It is a low metalloid steel with no severe restriction on residual content. It is used for moderate acid foods.

Type MC: It is similar to the MR type but has high phosphorus content to give mechanical strength or stiffness. It is usually used for low-acid foods.

Lacquering

To avoid corrosion and prevent black staining of the tin plate, it is necessary to coat the inside of the can with some material like lacquer. The process of coating the inside of a tin can is called lacquering. This is required to prevent discoloration and wholesomeness of the contents. Lacquers include oleo-resinous material, synthetic resins, phenolic resins, epoxy resins, and vinyl resins. There are two types of lacquers: (a) acid-resistant and (b) sulfur resistant. The acid-resistant lacquer is ordinary gold coloured enamel and the cans treated with it are called as A.R-enamel cans. The Sulphur-resistant lacquer is also of golden colour and the cans coated with it are called C-enamel cans or S.R. cans. Acid-resistant cans are used for packing of fruits of the acid group with soluble colouring matter such as raspberry, strawberry, red plum, coloured grapes, etc. Sulphur-resistant cans are used for non-acid products like peas, corn, beans, etc.

Syruping and Brining

For enhancing the flavour and taste and preventing excessive sourness of the acidified product, syrups are added to fruits and brine is added to the vegetables. The purpose of adding syrup or brine is to fill the gaps between the pieces of canned products and convenience of heat transfer during sterilization. Brine containing 1 to 2 percent of table salt is used for vegetables and for fruits, cane sugar, glucose syrup, invert sugar and high fructose corn syrups are added. Strength of syrup is measured by using hydrometer or a refractometer while strength of brine is measured by salometer or salinometer. The syrup or brine should be added to the can at a temperature of about 90^0C, leaving suitable headspace in the can.

Exhausting

Before sealing, heating the can and its content, is called exhausting. It is performed to remove air present in the interior of can and prevent corrosion. The vacuum created reduces undue strains upon the can during sterilization and minimizes oxygen remaining in the headspace result in the retention of Vitamins such as Vitamin C. it improves the shelf-life and also prevents bulging of the can when stored at high altitude or in hot climates. In exhausting, the can are passed through a tank of hot water at 92-97^0C or on a moving belt through a covered steam tunnel for 5 to 25 minutes. Vacuum also may be produced mechanically by sealing containers in a chamber under a high vacuum. It is followed by immediate sealing with the help of double rolling operation seamer.

Sealing: In sealing lids on metal cans, a double seam is created by interlocking the curl of the lid and flange of the can. Many closing machines are equipped to create vacuum in the headspace either mechanically or by steam-flow before lids are sealed.

Heat Sterilization

Heating of canned foods to destroy or inactivate bacteria is called as processing or retorting. Processing consists of determining just the temperature and the extent of cooking that would suffice to eliminate all possibilities of bacterial growth. In retort, saturated steam is supplied to heat the product. Time-temperature combination of processing depends upon the type and physical state of the product, the heat resistance of microorganisms or enzymes likely to be present in the food, the heating conditions, the pH of the food, and the size of the can to get complete sterility. In foods with higher pH ((pH

> 4.5); Clostridium botulinum is the most dangerous heat-resistant spore-forming pathogen likely to be present. Inside a sealed can, under anaerobic conditions, it can grow and produce a very fatal exotoxin, botulin. In more acidic foods (pH 4.5- 3.7), other microorganisms (e.g. yeast and fungi) or heat-resistant enzymes are used to establish processing times and temperatures. In acidic foods (pH < 3.7), enzyme inactivation is the main reason for processing and hence, heating conditions are less severe. The preservative effect of heat processing is due to the denaturation of proteins, which destroys enzyme activity and enzyme-controlled metabolism in microorganisms. The rate of destruction is a first-order reaction; that is when food is heated to a temperature that is high enough to destroy contaminating microorganisms, the same percentage die in each time interval regardless of the number present initially. This is known as the logarithmic order of death and is described by thermal death rate curve. The time needed to destroy 90% of the microorganisms (to reduce their numbers by a factor of 10) is referred to as the decimal reduction time or D-value. D-values differ for different microbial species (Table 13) and a higher D-value indicates greater resistance. The thermal destruction of microorganisms is temperature dependent and cells die more rapidly at higher temperature. By collating D-values at different temperatures, a thermal death time (TDT) curve is constructed.

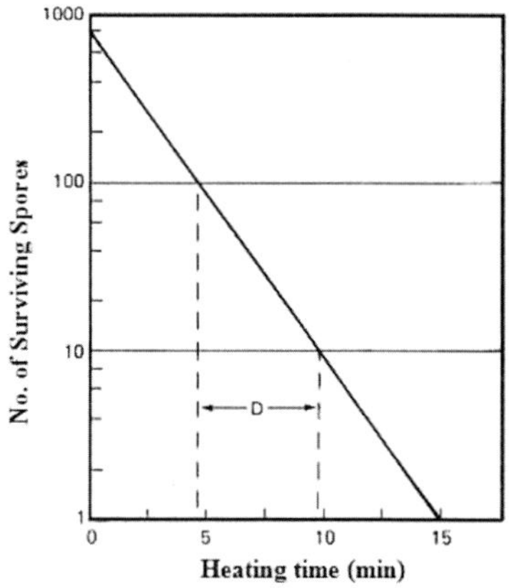

Figure 13: Thermal death rate curve

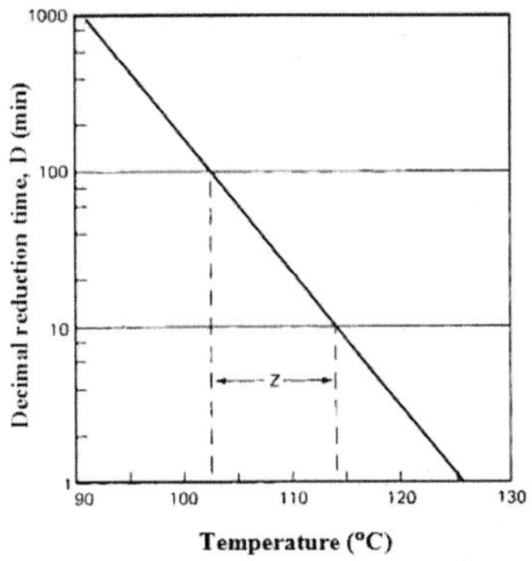

Figure 14: Thermal death time curve

The thermal death time or F-value is used as a basis for comparing heat sterilization procedures. F-value is the time required to achieve a specified reduction in microbial numbers at a given temperature and it represents the total time-temperature combination received by a food. The slope of the TDT curve is termed the z-value and is defined as the number of degrees Celsius required to bring about a 10-fold change in decimal reduction time. F-value is quoted with suffixes indicating the retort temperature and the z value of the target microorganism. For example, a process operating at 110^0 C based on a microorganism with a z-value of 10^0 C would be expressed as F^{10}_{110}. Hence, D-value and z-value are used to characterize the heat resistance of a microorganism and its temperature dependence, respectively while F-value is used for comparing sterilizing procedures.

Table 4: Heat resistance of some spore-forming bacteria used as a basis for heat sterilization of low-acid foods

Microorganism	z-value(°C)	D_{121} value(min)	Typicalfoods
Thermophilic (35-55°C)			
Bacillus stearothermophilus	9-10	3.0-4.0	Vegetables,milk
Clostridium thermosaccharolyticum	7.2-10	3.0-4.0	Vegetables
Mesophilic (10-40°C)			
Clostridium sporogenes	9.9-11.1	0.7-1.5	Meats
Bacillus subtilis	4.1-7.2	0.3-0.76	Milk products
Clostridium botulinumtoxins A and B	5.5	0.1-0.3	Low-acidfoods
Bacillus coagulans	6.9	0.01-0.07	Milk
Bacillus cereus	3.6	3.9	Milk
Psychrophilic (-5 to -1.5°C)	10	3.0 (60°C)	Low-acidfoods

Rate of Heat Penetration in Processing

Heat is transferred from steam or pressurized water through the container and into the fruit or vegetable or any other canned food. The heat transfer patterns in canned foods are given in Fig. 9.4. The rate of heat penetration is measured by placing a thermocouple at the thermal center of a container (the point of slowest heating) to record temperatures in the food during processing. It is assumed that all other points in the container receive more heat and are therefore adequately processed. The zone of slowest heating in a container is called as cold point, which is most difficult to sterilize. In cylindrical containers the thermal center is at the geometric centre for conductive heating foods and approximately one third up from base of the container for convective heating foods.

The important factors that influence the rate of heat penetration into a food are given below:

Type of product- Liquid or particulate foods (for example peas in brine) in which natural convection currents are established heat transfers faster than in solid food in which heat is transferred by conduction (for example pastes or purees). The low thermal conductivity of foods is a major limitation to heat transfer in conduction.

Size of the container- Heat penetration to the center is faster in small containers than in large containers.

Agitation of the container- End-over-end agitation and to a lesser extent, axial agitation increases the effectiveness of natural convection currents and thereby increases the rate of heat penetration in viscous or semi-solid foods (for example beans in tomato sauce).

Temperature of the retort- A higher temperature difference between the food and the heating medium causes faster heat penetration.

Shape of the container- Tall containers promote convection currents in convective heating.

Type of container- Heat penetration is faster through metal than through glass or plastics owing to differences in their thermal conductivity.

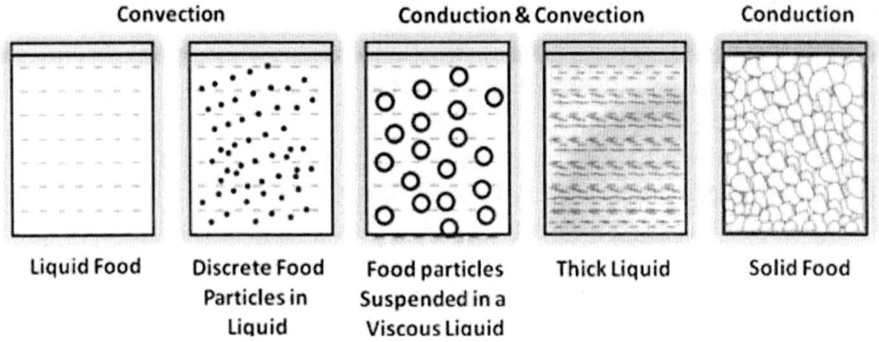

Figure 15: Heat transfer pattern in canned foods

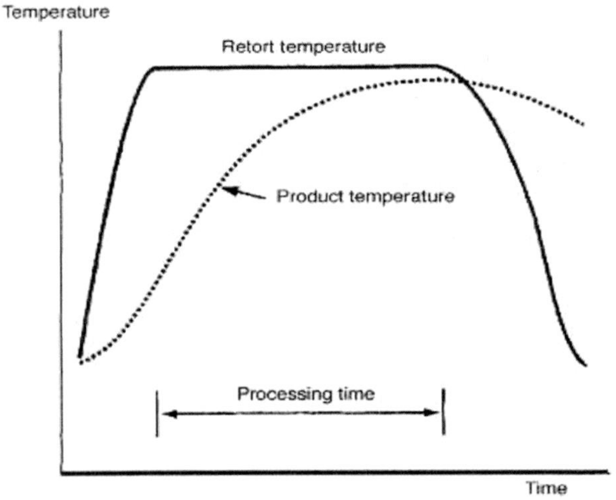

Figure 16: Heat penetration into a can of conductive heating food

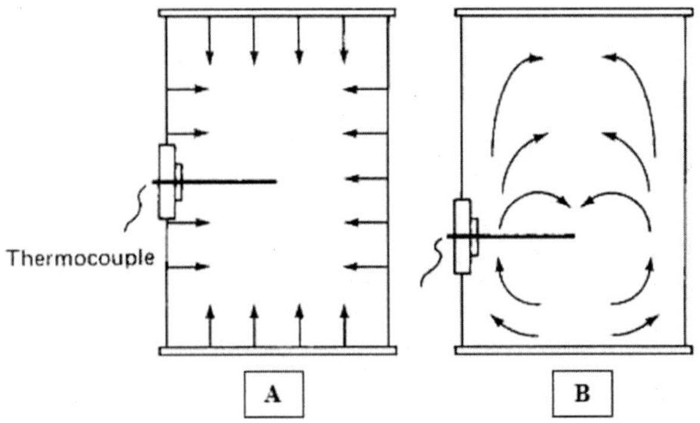

Figure 17: Heat transfer into containers by conduction **(A)** and convection **(B)**

Cooling – After heat sterilization, containers are quickly cooled to prevent overcooking. Containers may be quick cooled by adding water to the cooker under air pressure or by conveying the containers from the cooker to a rotary cooler equipped with a cold- water spray.

Labeling and Casing – After the heat sterilization, cooling, and drying operations, the containers are ready for labeling. Labeling machines apply glue and labels in one high-speed operation. The labeled cans or jars are the packed into shipping cartons. After labeling, cans are kept in wooden boxes or corrugated carboard cartons and stored in cool and dry place.

Canning methods

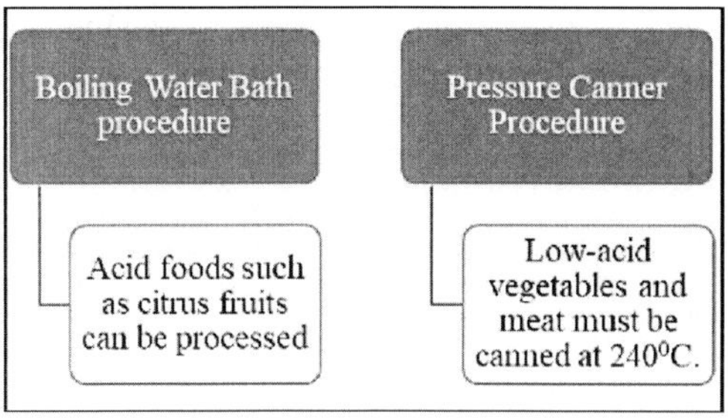

Boiling water Bath Procedure

1. Canner should be about half full of clean, hot water.
2. Load filled jars, one at a time, into the canner using a jar- lifter. Place them on a rack. Make sure the jar-lifter is securely positioned below the neck of the jar. Keep the jar upright at all times.
3. Add more boiling water, if needed, so that water level is 1 to 2 inches above the tops of the jars. Pour water around the jars, not on them to prevent breakage.
4. Turn the heat setting to the highest position, cover the canner with its lid, and heat until the water boils vigorously.
5. Set a timer after the water is boiling for the total minutes required to process the food.
6. Keep canner covered during the processing. The heat setting may be lowered as long as the water boils steadily for the entire time.
7. Add boiling water, if needed, to keep water above the jar tops.
8. If water stops boiling at any time during the process, turn the heat to its highest setting, bring the water back to a vigorous boil, and begin timing the process again, from the beginning (using original total processing time).

9. When the jars have been processed, turn off the heat and remove the canner lid. Wait 5 minutes before removing jars.
10. Using a jar-lifter, remove the jars one at a time. Do not tilt jars Carefully place hot jars right-side up directly onto dry towels or a cake cooling rack to prevent jars from breaking on contact with a cold surface. Leave at least 1 inch between the jars during cooling. Avoid placing the jars in a cold draft. Do not cover with towels.
11. Let the jars sit undisturbed while they cool for 12 to 24 hours. Do not tighten ring bands on the lids or push down on the center of the flat metal lid until jar has completely cooled.

Pressure Canner Procedure

1. Place the center canner over the burner. When the jars of food are ready for canning, put the rack and 2 to 3 inches of hot water into the canner. Begin heating the water, but not enough for the depth to decrease.
2. Use a jar-lifter to place filled jars, fitted with lids, on the jar rack in the canner. Leave space between the jars for steam to flow around each one during processing. Keep the jar upright at all times.
3. Fasten the canner lid securely. Leave the weight off the vent port or open the petcock.
4. Turn the heat setting to its highest position. Heat until the water boils and steam flows freely in a funnel shape from the open vent port or petcock. Let the steam flow continuously for 10 minutes (to vent the canner).
5. After venting the canner, place the counterweight or weighted gauge on the vent port or close the petcock, depending on the type of canner. If you have a weighted gauge, be sure to use the correct setting. The canner will begin to pressurize.
6. For a dial gauge canner, let the pressure rise quickly to 8 pounds of pressure. Lower the burner temperature slightly and let the pressure continue to rise to the correct setting.
7. For weighted gauge canners, let the canner heat quickly until steam begins to escape from the gauge or the gauge moves and makes noise. Adjust the heat down slightly until the weighted gauge begins to jiggle or rock as the manufacturer describes.
8. Start timing the process when the pressure reading on the dial gauge indicates that the recommended pressure has been reached or when the weighted gauge begins to jiggle or rock as described.

9. Adjust the heat under the canner to maintain a steady pressure at, or slightly above, the correct gauge pressure. If the pressure goes too high, do not lower it by opening the vent or lifting the weight. Instead, turn down the heat under the canner.
10. If at any time the pressure goes below the recommended amount, bring the canner back to pressure and begin timing the process again, from the beginning (using the total original process time).
11. When the time process is complete, turn off the heat, remove the canner from an electric burner if possible, and let the canner cool down naturally. If the canner is too heavy, simply turn off the heat.
12. Do not force-cool the canner or depressurize it. It will depressurize while it is cooling.
13. After the canner is depressurized, remove the weight from the vent port or open the petcock. After 10 minutes, unfasten the lid and remove it carefully. Lift the lid with the underside away from you so that the steam does not burn your face.
14. Do not leave the canner unopened to cool completely.
15. Using a jar-lifter, remove the jars, being careful not to tilt them. Carefully place the hot jars directly onto dry towels or a cooling rack. Leave at least 1 inch of space between jars.
16. Let the jars sit undisturbed while they cool for 12 to 24 hours. Do not tighten ring bands on the lids or push down on the center of the flat metal lid until the jar is completely cool.

The goal of the canning process is to destroy any microorganisms in the food and prevent recontamination by microorganisms. Heat is the most common agent used to destroy microorganisms. Removal of oxygen can be used in conjunction with other methods to prevent the growth of oxygen-requiring microorganisms.

3

Preparation, Preservation and FSSAI Specifications of Juices, Squashes, Syrups, Sherbets, Nectars, Cordials etc.

Juice and juice products represent a very important segment of the total processed fruit industry. Juice products are being marketed as refrigerated, shelf-stable, and frozen, in a variety of packages with increased emphasis on functionality, health attributes, new flavors or blends, and in some cases fortified with vitamins and minerals. High-quality juice operations are dependent upon a source of high- quality raw material.

Fruits provide an important source of energy for human beings. But the main drawback is that the availability of majority of fruits is seasonal and they are perishable. Hence, fruits need to be processed and preserved which extends their shelf life and also results in value-addition. India is a producer of many varieties of citrus fruits. Hence, it is sensible if processing industries are set up near the places where the fruits are grown. States like Maharashtra, Uttaranchal, Himachal Pradesh and the North-Eastern states are high producers of many fruit verities. Assam is a major producer of fruits such as pineapple, orange, lemon, mango etc. and it is grown in large quantities over here. Hence, fruit squash and cordial making activity in a small scale is highly recommended in these regions.

Fruit Juices

Most fruit juices are excellent sources of vitamin C, several are good sources of carotene and many contain moderate amounts of pyridoxine, inositol, folic acid and biotin. Fruit juice is regarded as source of energy due to their rich carbohydrate content. The organic acids present in the fruit juice plays a significant role in the maintenance of the acid-base balance in the body.

The process starts with sound fruit, freshly harvested from the field or taken from refrigerated or frozen storage. Thorough washing is usually necessary to remove dirt and foreign objects and may be followed by a sanitation step to decrease the load of contaminants. Sorting to remove decayed and moldy fruit is necessary to make sure that the final juice will not have a high microbial load, undesirable flavors, or mycotoxin contamination. For most fruits, preparation steps such as pitting and grinding is required prior to juice extraction. Heating and addition of enzymes might also be included before the mash is transferred to the extraction stage. Juice extraction can be performed by pressing or by enzymatic treatment followed by decanting. The extracted juice will then be treated according to the characteristics of the final product.

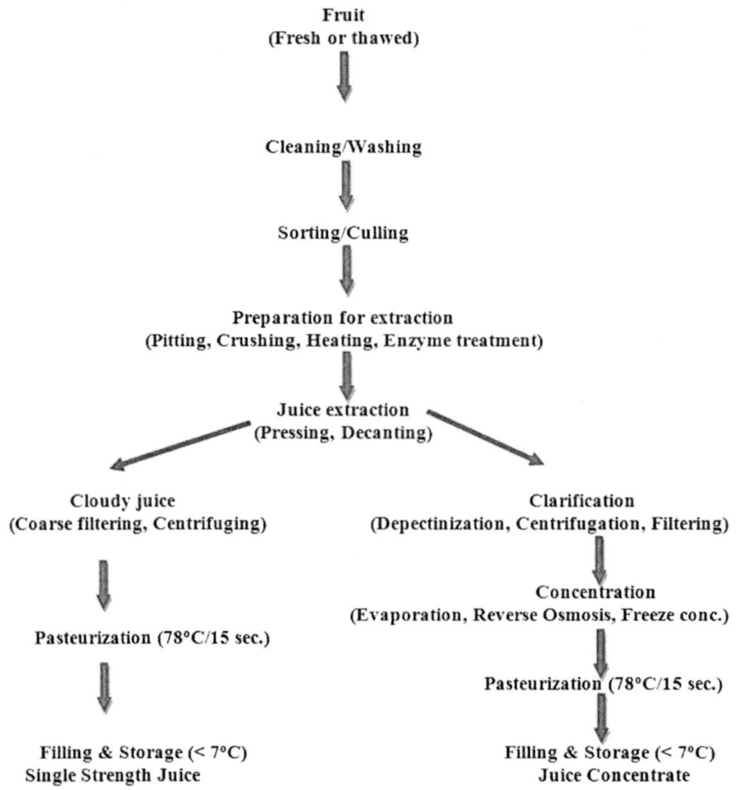

Figure 18: Generalized flow chart for fruit juice production

For cloudy juices, further clarification might not be necessary or may involve a coarse filtration or a controlled centrifugation to remove large insoluble particles. For clear juices, complete de- picturization by addition of enzymes, fine filtration, or high speed centrifugation is required to achieve visual clarity. The next step is usually a heat treatment or equivalent non-thermal process to achieve a safe and stable juice and final packaging if single- strength juice is being produced. For a concentrate, the juice is fed to an evaporator to remove water until the desired concentration level is obtained. Other processes used for water removal include reverse osmosis and freeze concentration, which are best suited for heat-sensitive juices. The concentrate is then ready for final processing, packaging, and storage.

Significance of Few Important Steps in Fruit Juice Production

Straining/Filtration/Clarification

- Extracted fruit juice contains varying number of suspended solids broken fruit tissue, seed, skin & various gums, pectic substances and proteins in colloidal suspension.
- Coarse particles removed by straining (non-corrodible metallic screens) or sedimentation.
- If clear juice required (grape, apple, lime juice cordial) complete removal of all suspensions effected through filtration or clarification with the help of fining agents and enzymes.
- Fining agents (i) Enzymes (pectolytic, starch liquefying, proteolytic), (ii) Mechanical finings (Infusorial earth), (iii) Chemical finings (Gelatin, albumin, casein).
- Apple juice 0.5-1.5 oz. of tannin and 1.5-6.0 oz. of gelatin per 100 gal. of juice fining purpose.

Disintegration: Juicing process starts with crushing, a step to break down the cell tissue. This may be accomplished using various type of mills viz., hammer mill, grinding disk mill, grating mill, crushers, stoned fruit mill or even turbo extractors.

Hot Break Process

In order to maximize juice yield and color-flavour extraction, a hot break process is often used.

Mash Enzyme Treatment

This step might not be used for the production of high quality, single-strength, cloudy and clear juices, where the preservation of the fresh flavor is imperative. Depectinization is designed to reduce the viscosity and slipperiness of the pulp and thus permit the effective use of decanters and presses with proper press aids as needed. It is especially useful in processing mature and stored fruit that results in low juice yield.

Fruit juice Extraction Process

Hydraulic rack and frame press is the most common batch press system used in small scale operations. Other type of extractors include horizontal piston press, bladder press, belt press, screw press or even decanter centrifuge.

Deaeration

Pure orange juice which is extremely susceptible to the adverse action of the residual air, is subjected, immediately after extraction, to a high vacuum whereby most of the air as well as other gases are removed.

Significance of Enzymes in Fruit Juice Production

Pectinases, cellulases and hemicellulases are used for clarification of fruit juices, juice extraction, improvement of cloud stability of vegetable and fruit juices and nectars, liquefaction and maceration of fruits and vegetables, reduction of cooking time of pulses and improvement of rehydration characteristics of dried vegetables.

These enzymes degrade structural polysaccharides that interfere with juice extraction, filtration, clarification and concentration. These enzymes are termed macerating or mash enzymes that finds application in extraction of juice from citrus fruits and from tropical fruits such as mango, banana and papaya and pineapple. For Apple and Pears, pectic enzymes are used to facilitate pressing or juice extraction, to aid in separation of flocculent precipitate by sedimentation, filtration or centrifugation.

Combination of pectinolytic enzymes are added to fruit and pressed fruit juice to reduce juice viscosity. Such viscosity reduction makes juice filtration, clarification and concentration more efficient. In concentrated fruit systems, it improves the efficiency of spray drying.

Table 5: Roles of enzymes in juice production

Enzymes	Function in juice production
pectinase enzymes	• Pectins are colloidal in nature, making solutions viscous and holding other materials in suspension. Pectinesterase removes methyl groups from the pectin molecules exposing carboxyl groups which in the presence of bi- or multivalent cations, such as calcium, form insoluble salts which can readily be removed. • Polygalacturonase degrades macromolecular pectin, causing reduction in viscosity and destroying the protective colloidal action so that suspended materialswill settle out. • Addition of pectic enzymes to grapes or other fruits during crushing or grinding results in increased yields of juice on pressing. Wine from grapes so treated will usually clear faster when fermentation is complete, and have better color. • Pectic enzymes are necessary for making high density fruit juice concentrates or purees. If apple juice is concentrated to 72°Brix without removal of the naturallyoccurring pectin, a gel will result rather than the desired liquid concentrate. • In most cases, juices are depectinized andfiltered before concentration, but in others the pectinase is allowed to act while the juice is being concentrated.
Cellulase	• Commercial cellulase products capable of hydrolyzing non-pectin polysaccharides such as cellulose, glucans, and xylans are used to fractionate the cell walls and liquefy the remaining solids.
Naringinase and Limoninase	• Used to hydrolyze naringin and limoninthe bitter compounds that are found in grapefruit juice.
Amylase and Arabinase	• Starch and araban imparts a cloudy appearance called haze, when released into the juice from certain fruits. Hazes due to starch are common in juice from early season apples and can be degraded by amylases. • Although araban is not a problem in extracted juice, it produces a permanent haze in concentrated juices, where it has limited solubility. Arabinase can be used tohydrolyze araban so that haze does not develop during storage of juice concentrates.

Benefits in the use of Enzymes

- Increased juice yield
- Improved efficiency of juice filtration
- Improved juice stability and concentration
- Enhanced juice clarity
- Reduced juice bitterness

Problems Encountered During Fruit Juice Processing

Some of the problems that may be faced in preparation of fruit juices are

- Browning of juice due to action of enzyme Poly phenol oxidase inherent in fruit (i.e. Apples).
- Foaming of fruit juice that may lead to oxidation of vitmin C and pose problem during filling of juice in packages.
- Bitterness of juice due to action of enzyme limonin and naringinase (e.g. in Orange).
- Cloudy juices due to pectinaceous substance (apple juice) or even bitartrates (i.e. argol precipitation) (i.e. in grape juice).

Table 6: FSSAI specifications for various fruit juices

Juice type	TSS (%), Min.	Acidity expressed as citric acid (%), Max.	Nutritive Sweet-ener (g/kg), Max.
Apple juice	10	3.5 (as)malicacid	-
Orange juice · Freshly expressed · Reconstituted	10	3.5	50
from concentrate	10	3.5	
Grape fruit juice	9		50
Lemon juice	6	4.0 Min	200
Lime juice		5.0 Min	200
Grape juice · Freshly expressed · Reconstitutedfrom concentrate	15 15	3.5 3.5	-
Pineapple juice · Freshly expressed · Reconstitutedfrom concentrate	10 10	3.5 3.5	50 -
Black currant	11	3.5	200
Mango, guava or anyother pulp fruit	15	3.5	GMP
Other fruit juice ofsingle species (not very acidic)	10	3.5	50
Other fruit juices ofsingle species (veryacidic)	10	3.5	200
Other fruit juices ofsingle species or combination thereof(not very acidic)	10	3.5	50
Other fruit juices ofsingle species or combination thereof(very acidic)	10	3.5	200

Squash

Squash is often colloquially known as "juice", especially when talking to young children because they might not be familiar with the term "squash". But this term is a misnomer as no squash is a pure juice. Squashes are commonly called by the name of the type of the fruit out of which they are made. Very rarely, they might be referred to as a "fruit drink", especially if they are ready diluted in a plastic bottle or paper carton

Squash is a type of fruit beverage which contains at least 25 per cent of fruit portion (juice/pulp) and not less than 40 per cent of total soluble solids (sugar). It also contains edible acid in the range of 1-2 per cent. Since the sugar content in squash is quite less therefore to preserve it, approved chemical preservative such as sodium or potassium metabisulphite or sodium benzoate should be added along with colour and flavour. Squash can be prepared from fruits such as mango, orange, pineapple, litchi, lemon etc.

"High Juice" is a squash that contains a larger amount of fruit juice. Up to half or more of the volume in juice is sold in markets as high juice. Many squashes contain less than 20 per cent juice and some squashes contain as less as 5 to 10 per cent juice. The latter have a lower nutritive content in contrast with the high juice variant which are reasonably higher in nutrients. Although one downside is that it is high in sugar and does not contain fiber or minor nutrients. That goes with almost all squashes. A low juice squash may state "with real fruit juice" on the label.

Theory

Squash should contain at least 25 per cent fruit juice or pulp and 40 to 50 per cent total soluble solids commercially. About 1 per cent citric acid and 350 ppm Sulphur dioxide or 600 ppm sodium benzoate are added as preservatives. By careful attention to hygienic condition the concentration of SO_2 in the squash can be safely reduced to about 250 ppm. Fruits most commonly used for preparing beverages are sweet orange, sour lime, mandarin, loose jacket orange, mango, apple, grape fruit, pineapple, lemon etc.

Method of Calculation

For the preparation of 10 liters of squash follow the procedure given below:

Calculate the amount of juice required as per commercial specification

Required juice = (25/100) × 10 = 2.5 L.

Measure the TSS using a refractometer (say the TSS is 30%)

Calculate the total solids content of the juice i.e. 0.3 X 2.5 = 0.75 kg

The final required TSS content in the product is to be saying 50 per cent. The TSS required to be added to obtain the final product is $(0.5 \times 10 - 0.75)$ kg = 4.25 kg.

The number of soluble solids in the form of citric acid and KMS is Citric acid @ 1 per cent, in the final produce is 100 g i.e. 0.1 kg. 600 ppm SO_2 (1.5g/litre of KMS being equivalent to 1000 ppm) @ 0.9g KMS/litre. i.e. 0.9 x 10 = 9 g i.e 0.009 kg.

The Number of solids to be added in the form of sugar is 4.25– (0.1+0.009) = 4.141 kg.

Preparation Procedure

1. As prescribe dissolve sugar in water, add citric acid and give a boil.
2. Any dirt is strain through a fine muslin cloth.
3. The syrup is cooled slightly.
4. Extract juice from fresh fruits by crushing and pressing them by using suitable juice extractors, basket presses or fruit pulper. Fruits, which require preheating, should be preheated before extraction.
5. Strain and filter the juice to remove suspended matter consisting of broken fruit tissue, seed, skin, etc. Clarify the juice if required using a suitable method.
6. The clean syrup is blended with fruit juice.
7. Add colour as required and then essence.
8. To improve flavour peal emulsion of 2 to 4 oranges for every 100 oranges taken or an appropriate quantity of an essential oil is added to the squash.
9. After mixing all the ingredients a calculated quantity of preservative, about 28 g for every 454 gram of squash is added.
10. Now fill the squash into sterilized bottles, leaving about 1.2 to 2.5 cm of head space.

Precautions

All equipment used in the preparation of fruit juices and beverages should be rust and acid proof.

Copper and iron vessels should be strictly avoided as these metals react with fruit acids, and cause blackening of the product.

Avoid exposure of juice to atmosphere as it will spoil the colour, taste and aroma and also reduce the Vitamin content.

Do not heat the juice.

- Do not mix hot sugar syrup in juice' or pulp.
- Do not use sodium or potassium meta bisulphite for coloured squashes.
- Always use stainless steel utensils, knives etc.

Table 7: FPO specification for Squash

Juice content	Not less than 25%
Total soluble solids	Not less than 40%
Acidity	Not more than 3.5%
Preservatives	Sulphur dioxide (not more than 350 ppm) Benzoic acid (not more than 600 ppm)
Synthetic sweeteningagent	Not permitted
Fermentation test	Negative at 37⁰C
Organoleptic quality	Free from objectionable taints and flavour

Cordials

Cordial is a sparkling, clear, sweetened fruit beverage from which pulp has been removed completely. It contains at least 25 per cent fruit juice, 30 per cent total soluble solids (TSS), 1.5 per cent acid and 350 ppm sulphur dioxide. Lemon and lime juice cordials are the commercial products.

Cordial, diluted juice and squash are more so similar products. But the major difference is that cordials tend to be thicker and stronger, requiring less syrup and more water to be blended. "High juice" is not a brand of squash but rather a type that contains a larger amount of juice, up to about 45 per cent.

Squashes are made out of fruit juice mixed with sugar syrup. Cordials on the other hand are crystal clear squashes. The process involves producing juice, which is then filtered through fine cloth or special juice filters to make the consistency crystal clear for cordials. Sugar syrup in the concentration range of 50 – 60 per cent is filtered and heated to 90⁰C and mixed in the exact proportion with the fruit juice. Addition of hot sugar syrup to the juice decreases the time that the juice needs to be heated and the colour and flavor of the juice are better preserved this way. This method also proves to be beneficial in order to reduce the equipment costs as sugar syrup can be heated using large aluminium pan, which is much more cost effective i.e. cheaper than using stainless steel. A small stainless steel pan is then used to finish heating the juice or syrup mixture. A very simple and easy means of calculating the amount of sugar and juice that should be mixed together (also when making jams) is to use a Pearson square. To employ this method, first a square is drawn. The juice and syrup concentrations are written on the left side and the required product concentration is written in the middle, for example, say 15 per cent. By subtracting the smaller amount from the larger amount diagonally, the quantities that should be mixed together can be found out. These drinks are diluted to taste with water and are therefore used a little at a time, so they may usually contain a preservative such as sodium benzoate to prevent spoilage once the bottle is opened. Although some processors make use of food dyes, these are not necessary for most of the products.

The ingredients used in the preparation of squashes and cordials have evolved over the years. A traditional cordial contains the following ingredients:

- Sugar
- Juice or plant extract
- Some water

It may also contain an acidifier such as citric acid. In very old- fashioned cordials lemon juice is added. Sometimes spices such as cinnamon or cloves may be added. Recreations of these traditional preparations often contain a preservative especially Sulphur dioxide. Sugar can also go a long way in keeping the squash or cordial preparation fresh for a long time.

Table 8: FPO specification for cordial

Juice content	Not less than 25%
Total soluble solids	Not less than 30%
Acidity	Not more than 3.5%
Preservatives	Sulphur dioxide (not more than 350 ppm) Benzoic acid (not more than 600 ppm)
Synthetic sweeteningagent	Not permitted
Clarity in cordial	Clear, free from pulp and other cellularmatter
Fermentation test	Negative at 37^{0}C
Organoleptic quality	Free from objectionable taints and flavour

Fruit Nectar

This is a type of fruit beverage containing at least 20 per cent fruit juice/pulp and 15 per cent total soluble solids besides about 0.3 per cent acid. It is also not diluted before serving.

Method of Preparation

For the preparation of 10L. of fruit nectar, follow the procedure given below:

- Calculate the amount of juice required as per commercial specification Required juice = (20/100) × 10 = 2.0 lts.
- Measure the TSS using a refractometer (say the TSS is 30%) Calculate the total solids content of the juice i.e. 0.3 × 2.0 = 0.60 kg.

- The final required TSS content in the product is to be say 15%. The TSS required to be added to obtain the final product is $(0.15 \times 10 - 0.60)$ kg = 0.90 kg.
- The number of soluble solids in the form of citric acid and KMS is Citric acid @ 0.3per cent, in the final produce is 30 g i.e. 0.03 kg.
- Number of solids to be added in the form of sugar is 0.90– 0.03 per cent = 0.87kg.
- Add calculated amount of sugar to about 2 L of water and heat it till it dissolves completely. Add citric acid and juice to the sugar syrup and makeup the volume to 10 L. Mix it well.
- Heat the RTSB / Nectar up to 90°C, fill hot in clean pre- sterilized glass bottle up to brim, seal and cool in the air or Fill in bottle, seal and heat process (90°C for 25 min.).

Sherbet

Sherbet should contain at least 65 per cent total soluble solids, suitably acidified and may or may not contain fruit juice.

Method of Preparation

For the preparation of 10 L. of sherbet follow the procedure given below:

- Take about 6 L. of water and add approx. 6.5 kg of sugar and dissolve it properly.
- Add the required flavour.
- Make up the volume to 10 L. and adjust the TSS to 65 per cent.

4

Equipment and Methods of Extraction and Clarification

Introduction

With the dynamic change in the pace of human activity and changing life style, the demand for easy-to-prepare drinks is also increasing. Besides being consumed in the household, these drinks are also served in restaurants, clubs, airlines, railways etc. There has been an appreciable increase in the export of processed foods which includes squashes, cordials, and syrups. With the fast pace of growth in the urban sector and the increase in living standards of the Indian people reaching higher levels by the year, there is a good potential to develop this industry in the small-scale sector. To prepare fruits and vegetable juices at commercial level, several equipments are needed. Some of the important equipment required for processing of specific fruit and vegetables are discussed herein.

Pome Fruit and Small Fruit Processing

Generally, most pome fruit (e.g. Apple, Pear, Quince) and small stone fruit (e.g. Plum, Olive, Peach, Cherry) can be used for juice extraction. No peeling is needed. Small stone fruit such as apricots and plums might have to be destoned (pitted) depending on the grinding, extraction equipment selection. Cherries, although containing a pit, may be pressed with the pit intact. Breakage of the pit will release benzaldehyde, the familiar aroma of maraschino- type cherries.

Disintegration

The juicing process starts with the crushing step to break down the cell tissue. Grain sizes of 5 - 8 mm diameter is recommended for presses, while grain sizes of 3 - 5 mm are desirable for decanters.

Hammer Mills

These are devices used to crush the whole fruit in preparation for pressing. Hammer mills consist of heavy stainless-steel bars spinning from a common axis under high-speed rotation. The fruit is disintegrated until it passes out through a screen of a specific size mounted in the bottom of the mill. With firm fruit, a small screen size should be used, and the mash will be of a finer particle size. Mash from firm fruit will press more easily, and the smaller particle size will allow greater yields.

Grinding Mills

They offer an alternative method to disintegrate fruit. The fruit was drawn past fixed knives mounted on a rotating cylinder. Control of the grind was accomplished by adjusting the depth of the knives and, thus, the size of the cut from the fruit.

Grinding Disk Mills

They offer more flexibility and improved performance. In the Bucher-Guyer unit, the fruit is transported by a feed screw to the grinding area. The screw pressurizes the fruit against a rotating disk equipped with grinding knives in a star pattern, and the milled fruit exits via an adjustable discharge slot. The process can be controlled by adjusting the feeder speed, the rotating speed of the grinding disk, the width of the product discharge slot (up to 10 mm), or by changing the knife size. Better yield is obtained by requisite adjustment corresponding to the fruit ripeness at the time of operation.

Grating Mills

These mills are used in small juice operations to produce uniformly sized fruit pieces. Fruit is fed to a rotating grating disk with fixed aperture, and the shredded fruit is discharged at the bottom. Fruit must be relatively firm with small seeds or pitted.

Stemmer/Crushers

These crushers are used in grape juice processing to remove residual stems, leaves, and petioles from grapes and to perform the initial crush of the fruit. These units are designed around a perforated rotating drum, with holes ~ 2.5 cm in diameter. While traversing the rotating drum, the grapes are caught by the perforated drum and knocked from the stems. Individual grapes are broken open or crushed in the process and dropped through the drum. Stems, leaves, etc. continue to the centre of the drum and are discharged at the end for waste. Grapes are generally put through the crusher to gently express the juice and free up the flesh, yet still not break the seeds. Breakage of the seeds releases increased amounts of phenolics, adding to the astringency of the juice.

Stoned Fruit Mills

Such mills are used for plums and apricots to crush the fruit without breaking the stones to avoid juice flavour changes and storage instability. Hard rubber-lobed wheels rotate simultaneously, forcing the fruit down and separating most of the flesh from the intact stone.

Turbo Extractors

These are used for extraction of juice and puree from fruits and vegetables. The cold extractor unit has a feeding section with a variable speed screw and a cutting head; a softening section consisting of a stator and rotor (roto-pulse); and an extraction area equipped with a rotor with paddles and a perforated cylindrical screen that continuously turns the product by centrifugal force. The extractor can be adjusted by changing the feeding speed, the rotor speed, the gap between the rotor and the screen, and the screen size. The fruit can be protected from oxidation by the injection of nitrogen gas or antioxidant solution to the cutting area through built-in openings.

Figure 19: Turbo extractor

Hot Break Process

To maximize juice yield and color-flavor extraction, a hot break process is often used. The most common use is in grape juice processing, but other fruits such as cherries, plums, and berries may also benefit. Increased interest in highly coloured juices, rich in phenolic compounds with associated health benefits, is driving the development of better techniques to preserve the functional components while maximizing the extraction. Typically, crushed fruit or mash passes through a large bore, tubular heat exchanger where it is heated to 50 to 60^{0}C. This stage, known as the *hot break process,* is designed to extract a large amount of color and assist in maximizing the yield. To the hot fruit, a pectolytic enzyme is added, and in case of red grape juice processing, kraft (wood pulp) paper is also added prior to pressing to serve as a press aid.

The addition of press aid to the mash provides coarseness and channels for the juice to exit. Alternative press aids include rice hulls, bleached kraft-fibre sheets or rolled stock, and ground wood pulp. If the juice is going to be extracted by decanting or centrifugation, then there is no need for press aids.

Mash Enzyme Treatment

This step might not be used for the production of high quality, single-strength, cloudy and clear juices, where the preservation of the fresh flavor is imperative. Soluble pectin found in fresh juice as a result of the activity of pectolytic enzymes that are located in the fruit cell wall. The soluble pectin is the cause for difficulty in juice extraction due to increased juice viscosity and the lubrication it affords the press cake. Typically, the fruit mash is heated to 45 to 50^{0}C followed by the addition of pectolytic enzymes. Reaction time can take up to 1 - 2 h.

De-picturization is designed to reduce the viscosity and slipperiness of the pulp and thus permit the effective use of decanters and presses with proper press aids. It is especially useful in processing mature and stored fruit that results in low juice yield. Several depeptidizing tanks are employed so that a continuous flow may be maintained to the presses or decanters. Treatment of the mash with enzymes is expected to increase the yield, reduce the processing time, and improve the extraction of valued components of the fruit.

Fruit Juice Extraction Equipment

Rack and Frame Hydraulic Press

The hydraulic rack and frame press is a very common batch press system found in small juice operations (Figure 20). Heavy cotton or nylon cloths are filled with a set amount of mash and then folded to produce what is called a *cheese.* The individual cheese is stacked and separated by a wooden, stainless steel, or plastic spacer platen. The combined stack is then compressed using a hydraulic ram, during which the juice is expressed. The process delivers good yield but is labor intensive.

Figure 20: Hydraulic rack and frame press

Horizontal piston press

One of the most successful press systems in the fruit juice market is the Bucher horizontal piston press, Switzerland. This press is capable of pressing berries, stone fruit, and vegetables. It operates in batch mode with loads of up to 14 t/ filling. Flexible drainage elements covered with a nylon filter cloth carry the expressed juice out to a manifold.

The Bucher-Guyer Press is a highly automated pressing system used in a batch pressing operation. Generally, this system consists of a rotatable basket or cylinder with a hydraulic ram used for juice expression. Within the cylinder are fabric-covered flexible rubber rods with longitudinal grooves in them, that allow the juice to transport easily to the discharge port.

Bladder Press

The Willmes Press is a commonly used system for grape juice pressing. It is a pneumatic-based system that consists of a perforated, rotatable, horizontal cylinder with an inflatable rubber tube (air bag) in the center. The cylinder is filled with grape mash through a door on the cylinder wall, which is rotated to the top position. After filling, the press is rotated to ensure even filling. During this rotation, the air bag is filled, creating the mash compression action. The bag is then collapsed, and the cylinder is rotated. The rotation and pneumatic compression of the mash is repeated many times with increasing air pressure.

Belt Press

The continuous belt press is effective for grape and apple juice processing. In belt presses, a layer of mash is pumped onto the belt entering the machine. Press aid may be added for improved yield and reduced suspended solids. The belt is either folded over or another belt is layered on top of the one carrying the mash. A series of pressurized rollers compress the enveloped mash. Expressed juice is caught in drip pans. The cake is discharged from the last pressure roller.

Screw Press

A typical screw press consists of a reinforced, stainless steel cylindrical screen enclosing a large bore screw with narrow clearance between the screw and the screen (Figure 12). Breaker bars are located between the screw intervals in order to disrupt the compressing mash. Back pressure is provided at the end of the chamber and is usually adjustable. Capacities for screw presses with diameters of 30.5 and 41.0 cm are 5,080 and 15,240 kg/h, respectively.

Figure 21: Conical screw press

Decanter Centrifuge

In addition to sieving technology, the separation of juice from the mash can be performed by sedimentation through increased gravity in a decanter. Centrifugal force is used to accelerate the settling of higher density insoluble particles present in the juice. Enzyme- treated mash is best suited for juicing by decanters, as the reduced viscosity and higher temperatures result in faster and more effective separation. The photograph of such decanter is furnished in Figure 22.

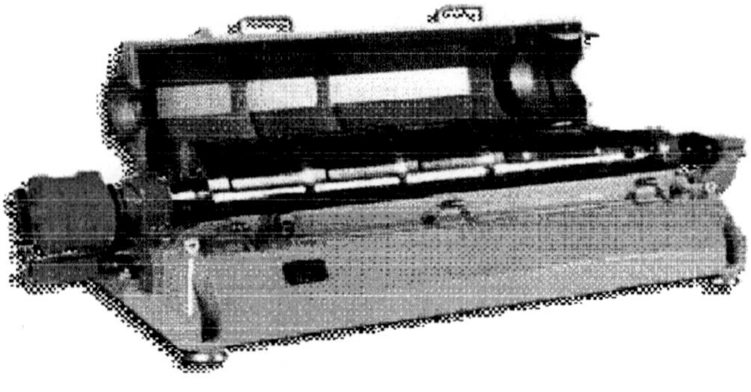

Figure 22: Decanter centrifuge

Pulper Cum Finisher

The separation of liquid and solids is accomplished by means of paddles rotating concentrically within a cylindrical screen. The liquid and desired amount of solid passes through the screen. The balance of the solids (pomace) is discharged through a large non- plugging port. The dryness of the pomace with a given screen can be controlled by paddle speed, pitch, clearance, or feed rate. Production throughput is dependent on the type of product being prepared, screen hole size and open area, paddle speed and pitch. An inlet impellor for breaking or macerating is available as an optional accessory.

Diverse type of materials viz. apricots, tomatoes, pumpkin, pears, apple, plums, berries, prunes and figs can be satisfactorily reduced to pulp, free of seeds, skins and fiber. Products such as citrus juices, jam, soup, peanut butter, jelly and fruit nectar can be finished to uniform clarification consistency. The picture of such machine is shown in Figure 23.

Figure 23: Pulper cum finisher

Mechanical Separation

For clarification of juice, after the enzyme (Pectinase) treatment, the sediment able particulates are separated by mechanical means. The equipment used for such process are as follows:

Decanters and Finishers

A high-solids stream can be partially clarified using decanters and finishers. Both pieces of equipment operate on the same principle with a spinning central cone, drum, and set of paddles pushing the juice through a screen of some type. The unit is typically mounted horizontally, and throughput is relatively high. Total suspended solids may be reduced to $< 1\%$ during operation, depending upon the characteristics of the feed stream and operating conditions of the separator.

Centrifugation

It is used for removal of juice-insoluble solids. A centrifuge places the juice under high gravimetric force induced by centrifugal action. This is effective in producing a juice that is opaque but free of visible solids. Modern centrifuges are highly automated and run continuously with timed solids ejection. Centrifuges with a high force of gravity are capable of producing clear juice under optimized conditions. Centrifuge must be operated in a manner to minimize the introduction of oxygen in the product. Possible remedies include the use of inert gas.

Pressure Filtration Devices

Filter Press

The cost is typically lower than other types of pressure filters. The system can be dismantled easily for inspection and cleaning. Filter cakes can be easily washed from the system once disassembling has progressed. In the filter press, the amount of unfiltered liquid is relatively low once the shutdown process is terminated.

Cylindrical Element Filter

In this system, tubular elements are suspended vertically in a closed tank system. Juice enters from the base of the system and filters through the elements, and the filtrate exits from the top of the system. Wash down and automation of this system are relatively straightforward.

Vertical leaf filter

It is a low-cost system because of the inherent simplicity of its design. It offers an easy cake removal system and can be automated. A modified version of the leaf filter is the horizontal tank vertical leaf filter that accommodates a very large area of filtration surface, up to 2000 ft^2 (180 m^2). Filter leaves can easily be removed, inspected, and repaired.

Rotating Leaf Filter

In this filter,

and rapid cake removal and clean-up system.

Horizontal Rotating Leaf Filter

It is essentially identical to the vertical rotating leaf system, except that it is available in much smaller models.

Clarification of Fruit Juice

Consumers have a strong preference for clear juices. In order to have attractive appearance of finished fruit juice, especially for beverages like fruit juice cordial, clarification of juice is highly essential. Such clarification can be done by the help of centrifugation and use of pectinase followed by decantation. Filter aids such as Infusorial earth, bentonite helps in achieving better clarification of fruit juices. Ultrafiltration (UF) and microfiltration (MF) have been used commercially for the clarification of fruit juices. After extraction, the fruit juice after de-pectinization is fed to UF unit for clarification. If the juice contains strong colour, microfiltration can be suitable for avoiding colour losses. Pre- centrifugation (10,000g for 15 min) of juice (especially cherry juice) before clarification is recommended.

Tropical countries, like India, have a vast scope of providing delicious cold drinks during hot summer particularly the fruit beverages. Due to increased consumer awareness with respect of quality, safety and health, these fruit beverages are becoming more and more popular and are gradually acquiring a chunk of the market share of cold drinks. Fruit beverages are easily digestible, highly refreshing, thirst quenching, appetizing and nutritionally far more superior to the synthetic aerated drinks.

Module-4

1

Preservation and Value Addition Techniques of Horticulture Crops

Definition: Food preservation can be defined as the science which deals with the process of prevention of decay or spoilage of food thus allowing it to be stored in a safe condition for future use. All foods gradually undergo deterioration or spoilage from the time they are harvested or manufactured. Some commodities spoil rapidly, while others can keep for longer but limited periods.

Preservation is a set of techniques used to extend the shelf life of perishable fruits and vegetables, maintaining their quality, nutrition, and safety. Proper preservation methods are crucial in post-harvest management to reduce wastage, meet consumer demands, and ensure food security. Let us explore the principles and methods of preservation using easy vocabulary

Principles of Preservation

- **Temperature Control**: Preservation involves maintaining low temperatures to slow down enzymatic and microbial activity. Lower temperatures can inhibit the growth of spoilage microorganisms and extend the freshness of fruits and vegetables.
- **Removal of Air and Moisture**: Air and moisture can accelerate deterioration and spoilage. Preservation methods often involve vacuum sealing or drying to remove air and reduce moisture content.
- **pH Regulation**: Adjusting the pH level can inhibit microbial growth. Acidic environments are unfavorable for many spoilage bacteria and fungi, enhancing the shelf life of preserved products.
- **Use of Preservatives**: Natural or chemical preservatives can be used to prevent spoilage and microbial growth. These compounds inhibit the activity of microorganisms and enzymes.
- **Packaging**: Proper packaging is essential in preserving fruits and vegetables. Packaging materials should be impermeable to air and moisture to maintain product quality and safety.

The Basic Purposes of food Preservation are

1. Extension of shelf-life of foods thus increasing the supply.
2. Ensuring the availability of seasonal foods throughout the year.
3. Adding variety to the diet.
4. Saving time by reducing preparation time and energy.
5. Stabilizing firm prices and prices of food.
6. Improving the nutritional qualities.

The food spoilage might be due to growth and activity of micro- organisms, insects or rodents, action of enzymes. Chemical reactions and physical changes take place during processing by burning, drying, freezing etc.

Methods of Preservation:

- **Refrigeration**: Refrigeration is a widely used method for short-term preservation. Storing fruits and vegetables at low temperatures slows down respiration and reduces spoilage. However, refrigeration is not suitable for all produce, as some items are sensitive to cold temperatures.
- **Freezing**: Freezing is an effective long-term preservation method. It involves rapidly reducing the temperature of fruits and vegetables to extremely low levels. Freezing halts enzymatic and microbial activity, preserving the quality and nutrients of produce.
- **Drying**: Drying or dehydration involves removing the water content from fruits and vegetables. This method inhibits microbial growth by reducing water availability, thus extending shelf life. Dried products are lightweight and have a longer storage life.
- **Canning**: Canning is a heat-based preservation method. Fruits and vegetables are sealed in airtight containers and heated to destroy microorganisms. This process ensures safety and allows long-term storage of produce.
- **Pickling**: Pickling involves immersing fruits and vegetables in acidic solutions, such as vinegar or brine. The acidic environment preserves the produce and enhances flavour.
- **Jam and Jelly Making**: Making jams and jellies involves cooking fruits with sugar and pectin. The high sugar content inhibits microbial growth and extends shelf life.
- **Fermentation**: Fermentation is a traditional preservation method that involves the breakdown of sugars by microorganisms. The by-products of fermentation, such as alcohol and acids, prevent spoilage.

- **High-Pressure Processing (HPP):** HPP involves subjecting packaged fruits and vegetables to high levels of hydrostatic pressure. This process inactivates microorganisms while retaining nutritional value and sensory qualities.

Preservation methods are chosen based on the type of produce, desired shelf life, available resources, and the intended use of the preserved products. Proper preservation practices are essential for minimizing post-harvest losses, ensuring food availability throughout the year, and adding value to agricultural products.

Value Addition Concept

Value addition is a strategic approach in post-harvest management that involves enhancing the quality, market value, and economic potential of agricultural products, such as fruits and vegetables. It aims to transform raw produce into value-added products with increased desirability and utility for consumers and markets. Value addition plays a crucial role in modern agriculture, food processing, and marketing. Let's explore the concept of value addition in post-harvest management

- **Diversification of Products**: Value addition involves the creation of a diverse range of products from raw fruits and vegetables. For instance, fresh fruits can be transformed into juices, jams, and dried snacks. Vegetables can be processed into pickles, canned products, or frozen items. These value-added products provide consumers with various options, increasing the demand and marketability of agricultural produce.
- **Quality Enhancement**: Value addition focuses on improving the quality of agricultural products. Through proper post-harvest handling, processing, and packaging, the nutritional value, appearance, taste, and shelf life of fruits and vegetables can be enhanced. Quality improvement boosts consumer satisfaction and builds trust in the brand or product.
- **Market Value and Price Enhancement**: Value-added products generally command higher prices in the market compared to raw produce. Processing and packaging transform fruits and vegetables into convenient, ready-to- use forms, leading to increased perceived value. As a result, farmers and producers can achieve better returns on their investments, leading to improved economic viability.
- **Waste Reduction and Profit Generation:** Value addition minimizes post-harvest losses by utilizing surplus or less visually appealing produce that might otherwise go to waste. By converting such produce

into processed goods or by-products, farmers can generate additional income and reduce overall waste.

- **Extending Shelf Life**: Value-added products often have an extended shelf life compared to raw fruits and vegetables. Through methods such as drying, freezing, canning, or pickling, the spoilage rate is significantly reduced, allowing products to be available for an extended period, even in the off-season.
- **Creating New Market Opportunities:** Value addition opens doors to new market opportunities and consumer segments. Processed fruits and vegetables can cater to diverse tastes, preferences, and dietary requirements, appealing to a wider audience and expanding market reach.
- **Promotion of Food Security**: Value addition helps preserve the nutritional value of fruits and vegetables. By processing and storing them in various forms, essential nutrients are retained and can be made available throughout the year. This contributes to food security, especially during times of scarcity or adverse weather conditions.
- **Employment Generation**: Value-addition activities require skilled labour for processing, packaging, marketing, and distribution. This leads to job creation, especially in rural areas, promoting economic development and improving livelihoods.

2

Preparation, Preservation, and Manufacturing Value - Added Products

Fruits and vegetables are one of the most consumable commodities globally and also serves as an important ingredient in human dietaries. There are a lot of different varieties of fruit and vegetable and the prices keep fluctuating of the products. The production of fruit and vegetables needs to be handled, processed, or utilized properly. The costs involved in preventing the losses are always cheaper than the cost of production, therefore processing receives higher attention in recent years. The fruits and vegetables processing industry was earlier highly unorganized but due to the increasing demand for processed food, the industry has started attracting investors. India facilitates the fruits and vegetables processing industry in obtaining its primary resources; India is the second-largest producer of fruits and vegetables. The global fruit and vegetable processing market includes the global fruit & vegetable processing equipment market and processed fruits & vegetables market. There are different methods included in fruits and vegetable processing that helps in increasing the life of seasonal fruits and vegetables are canning, dehydration, freezing, pickling, etc.

Crystallized Fruits

Candied fruit when covered or coated with crystals of sugar, either by rolling in finely powdered sugar or by allowing sugar crystals to deposit on them from dense syrup are called crystallized fruits.

In this case syrup of 70^0Brix is employed. It is placed in a large deep vessel and allowed to cool to room temperature. To avoid premature granulation of sugar, a sheet of waxed paper is placed on the surface of the syrup. The candied fruit is placed on a wire mesh tray, which in turn is placed in a deep vessel. The cooled syrup is then gently poured over the fruit so as to cover it entirely. To prevent the fruit from floating, another wire-mesh tray is placed on it and a sheet of waxed paper is placed on the surface of the syrup. The whole mass is left undisturbed for 12-18 h, at the end of which a thin crust of crystallized sugar is formed. The tray containing the fruit is then removed carefully from the pan, and the surplus syrup drained off. The drained fruit are spread in a single layer on wire mesh trays and allowed to dry at room temperature or in driers at about 49^{0}C.

Preparation

Cooled syrup (70% total soluble solids) is gently poured over the fruit so as to cover it entirely. The whole mass is left undisturbed for 12 to 18 hours during which a thin coating of crystallized sugar is formed. hours during which a thin coating of crystallized sugar is formed. The tray is then taken out carefully from the vessel and the surplus syrup drained off. The fruits are then placed in a single layer on wire mesh trays and dried at room temperature or at about 49^{0}C in driers.

Problems in Preparation of Preserves and Candied Fruits

Fermentation

Reasons: It is due to low concentration of sugar used in the initial stages of preparation of preserves. Sometimes fermentation also occurs during storage due to low concentration of sugar and insufficient cooking.

Prevention: Boiling the product at suitable intervals, by adding the required quantity of sugar and by storage in a cool and dry place.

Floating of fruits in jar

Reason: It is mainly due to filling the preserve without cooling. **Prevention**: Can be avoided by cooling the preserve prior to filling.

Toughening of fruit skin or peel

Reason: It may be due to inadequate blanching or cooking of fruits. Toughness may develop when cooking is done in a large shallow pan with only a small quantity of syrup.

Prevention: Blanching till tender is necessary

Fruit shrinkage

Reason: Cooking of fruits in heavy syrup greatly reduces absorption of sugar and causes shrinkage.

Stickiness: Fruits should be blanched first or cooked in low-sugar syrup.

Reason: It may develop after drying or during storage due to insufficient consistency of the syrup, poor quality packing and damp storage conditions.

Prevention: Good quality packing and storage under dry conditions required.

Fruit Preserves

Fruits are seasonal in nature. Further, it has got limited shelf life. Sugar has got the ability to increase the shelf life of fruit products through its osmotic pressure action on microorganisms. Preparation of fruit preserves is one novel method of preserving the seasonal fruit for use in non-season or to make available the 'fruit of choice' for the consumers to enjoy throughout the year. A mature fruit or its pieces impregnated with heavy sugar syrup till it becomes tender and transparent is known as a preserve. Apple, pear, mango, cherry, karonda, strawberry, pineapple, papaya, etc., can be used for making preserves.

Preserves (Murabbas) are prepared from whole fruits and vegetables or their segments by addition of sugar followed by evaporation to a point where microbial spoilage cannot occur. The final soluble solids concentration is reached to about 70 per cent.

The finished product can be stored without hermetic sealing and refrigeration. The fruit for preserving should be in a firm-ripe rather than a soft-ripe stage. By using up to 25.0 per cent of firm-ripe fruit, the tartness is increased and less pectin is required in the formula. The fruit should be uniform in size and uniform pieces to cook evenly. In preserves fruits can be cooked in syrup by three processes that are:

- Rapid process,
- Slow process and
- Vacuum process. Preserves made by vacuum process retain the flavour and colour of the fruit better other methods.

Most suitable fruits for preserves making are *Aonla, beal,* apple, pear, mango, cherry, *karonda*, strawberry, pineapple, papaya, peaches, plums, grapes, muscadines, quinces and tomatoes.

Processing

The fruit should be washed thoroughly. If the fruit had been sprayed with insecticide to check blight, it should be washed with dilute HCl, especially is the peel is not removed from the fruit for preparation of the preserve.

The preliminary treatment varies with the variety of fruit taken. For instance, apples and pears are just peeled and pricked if they are to be kept whole, otherwise they are peeled, halved or quartered, cored and punctured. Mangoes are peeled, sliced and pricked. Peaches are destoned and lye peeled. Apricots, cherries are only pitted and destined. Oranges, lemons, grape fruits and citrus fruits are halved and the pulpy portions removed from the cut fruit. Pumpkins are sliced, peeled, pricked and placed in dilute lime water for few hours to harden the texture. Strawberries and Raspberries are used as such, without any preliminary treatment.

The fruit is first cooked slightly in water to make it soft enough to absorb sugar. Cooking of the fruit in syrup is rather a difficult process, because the syrup is to be maintained at a proper degree of consistency so that it can permeate the entire body of the fruit, without causing it either to shrink or to toughen. The fruit should be boiled initially in water before transferring it to the syrup or cooked first in a dilute syrup with low sugar concentration. Highly juicy fruits may, however, be placed directly in a thick syrup at the very outset, because the excess of juice present in the fruit would serve to dilute the syrup.

Cooking in Syrup

There are three ways of cooking the fruit in a syrup. These are: (i) Open Kettle one period process, (ii) Open kettle slow process and (iii) vacuum cooking process.

Open Kettle one Period Process

The concentration of the starting sugar syrup for cooking the fruit should be low. Boiling should be continued with gentle heating until the syrup becomes sufficiently thick. Rapid boiling will make the fruit tough, especially when heating is done in large shallow pans with a small quantity of syrup. Soft fruits such as strawberries and raspberries require little boiling for softening, unlike hard fruits like apple, pear, peach, etc. which require prolonged heating. The final concentration of sugar syrup should not be less than 68°Brix, which corresponds to a boiling point of 106°C at sea level.

Open Kettle Slow Process

The fruit is first cooked in water to make it tender. Sugar equal to half the weight of fruit is added to the boiled fruit in alternate layers in a vessel and the mass is allowed to stand for 24 h. During this period, the fruit gives out excess water, and the sugar goes into solution resulting in a syrup of ~ 38°Brix. More sugar is added to raise the strength of the syrup to about 60°Brix. A small quantity of citric or tartaric acid (62-125 g per 100 kg of sugar initially taken) is added to invert a portion of the cane sugar. The whole mass is then boiled for 4-5 min and kept for 24 h. On the 3rd day, the strength of syrup is raised to ~ 68°Brix by adding sugar and the mass is boiled again for 4-5 min. The fruit is then left in the syrup for 3-4 days. Finally, the strength of the syrup is raised to 70°Brix and the preserve, in finished condition, is then packed in containers.

Vacuum Cooking

Preserves made by cooking under vacuum retain the flavor and colour better than those made in open kettles. In vacuum cooking process, the fruit is initially softened by boiling and then placed in the syrup, which should be about 30-35°Brix. The fruit syrup blend is then transferred to a vacuum pan and concentrated under reduced pressure to ~ 70°Brix. Hard fruits like apples and pears require slow boiling to facilitate penetration of sugar, while soft fruits can be boiled briskly.

Cooling and Packing

If the preserve is to be stored in bulk, it should be cooled quickly after the first boiling to avoid any dis-colouration of the product. The fruit is drained from the syrup and filled into dry containers. Freshly prepared boiling syrup of 68^0Brix is then poured into the containers such as in A 2.5 size cans, which are exhausted for 8-10 min at about 100^0C in steam, and then hermetically closed. If the preserve is packed scalding hot in dry cans, subsequent sterilization of the cans may be omitted. A 2.5 size (can hold 28.5 fluid ounce i.e. 843.6 ml) can may be preferably sterilized for about 25 min at 100^0C and cooled immediately thereafter.

If the preserve is to be candied, the Brix of the syrup should be raised to 75^0Brix and again allowed to stand for another week, before it is ready for candying.

Candied Fruits

The process is similar to the one employed in preparing fruit preserves, but here the fruit is impregnated with a higher percentage of sugar or glucose; the total sugar content is about 75 per cent. A certain amount of invert sugar or glucose is substituted in place of cane sugar.

The fruits suitable for such preparation are those that possess pronounced flavor, such as pineapple, peach, peels of orange, lemon, grapefruit, citrus, cherry, etc. Use of slightly underripe fruit helps in preventing formation of jam-like consistency in the syruping process.

The sweeteners used maybe Confectioner's glucose (corn syrup, crystal syrup or commercial glucose), dextrose, invert sugar, etc.

Jam

Introduction

Sweet spreads are foods with many textures, flavors and colors. They are thickened or jellied to varying degrees. The traditional jellies and jams are preserved primarily by sugar. The texture of jam is thick and mildly chunky, with small-ish bits of fruit. Crushed or cut-up fruits are cooked with sugar until soft and spreadable.

Jam is a type of IMF (intermediate moisture food) made by cooking crushed or chopped fruits with sugar to form a thick, spreadable consistency. The process involves heating the fruit- sugar mixture to a gel-like state, which sets upon cooling. The high sugar content acts as a preservative, inhibiting microbial growth and extending shelf life.

As per FSSAI, Jam means the product prepared from sound, ripe, fresh, dehydrated, frozen or previously packed fruits including fruit juices, fruit pulp, fruit juice concentrate or dry fruit by boiling its pieces or pulp or puree with nutritive sweeteners namely sugar, dextrose, invert sugar or liquid glucose to a suitable consistency. It may also contain fruit pieces and any other ingredients suitable to the products. It may be prepared from any of the suitable fruits, singly or in combination. It shall have the flavour of the original fruit(s) and shall be free from burnt or objectionable flavours and crystallization.

The Specific Requirements for Jam are as follows

Total soluble solids (m/m) - Min. 65 per cent.

The product shall be manufactured from minimum 45 per cent by weight, of original prepared, fruit, exclusive of any added sugar or optional ingredients of finished product, except for Strawberry or Raspberry fruit, where it shall contain minimum 25 per cent fruit.

Salient Characteristics of Jam

The amount of pectin in the gel is quite small (< 1.0% of the weight of jam); sugar content is very high (60-70%). Of the remainder, the bulk is water with a small amount of fibrous matter and seeds. For a chunky texture, some pieces of fruit are added about 5 minutes before the cooking is finished. If a stiffer or sweeter jam is desired, more thickener or pectin, honey or sugar may be added.

- **Fruit or fruit combinations:** Some fruits such as lemons and bitter oranges are rich in both acid and pectin and can be easily made into jam. Some popular combinations include: pineapple and paw paw, orange and banana, apple and plum.
- **Sugar:** 55 parts of sugar is required for every 45 parts of fruit. With excess sugar, the jam becomes gummy and sticky. The finished jam should have 30-50 per cent invert sugar / glucose to avoid crystallization of cane sugar during storage.
- **Acid**: Acidity can be supplemented by use of citric, tartaric or malic acid. The pH influences inversion of sugar and setting of the jam. For

pectin jam, pH of 3.3 gives a good set. For Apple and Plum jam, the pH range of 3.2-3.5 is desirable; optimum being 3.35.

Preparation of jam

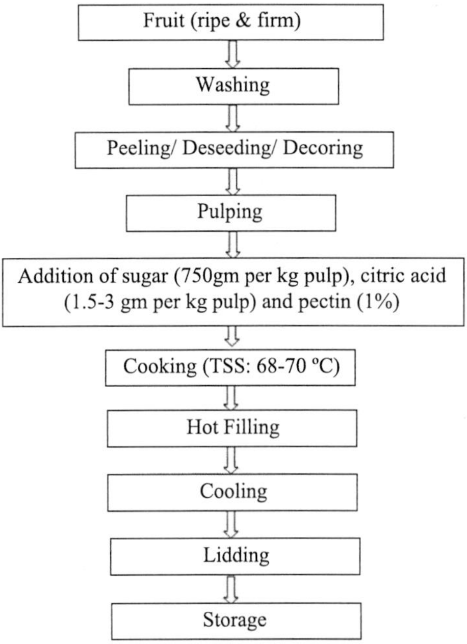

Figure 24: Flow chart for preparation of jam

Judging the End-point or Point at which Boiling is Stopped

Determining the Boiling Point with Thermometer

Thermometers or thermocouples are used to indicate the temperature which should be approximately the boiling point of a 65 per cent sugar solution (103.9-105°C), when the process is complete. 65 per cent solids boils at 104.8°C at sea level. It should be 8-9°C higher than the boiling point of water at that place.

- **Hydrometers**: They help in determining the specific gravity of material.
- **Refractometer**: We can determine the percentage solids i.e. TSS read directly on scale. Use Abe or Zeiss type refractometers and for dark-coloured jams/jelly, use projection type refractometer.
- Sheeting or ladle test

  D **Cold plate test**: A drop of boiling liquid from the pan is placed on a plate and allowed to cool. If the jelly is about to set, it will crinkle when pushed with a finger.

D **Sheet or Flake test**: Some portion of a jelly is taken in a large spoon or wooden ladle and cooled slightly. When dropped, if it falls in the form of flake or sheet, the end point is reached. If the jelly drips like syrup, it is required to further concentrate.

Weighing

The boiling pan is weighed before and after putting the fruit extract and sugar in it. The weight of finished jelly should be 1.5 times the weight of sugar used. Industry uses specific gravity method, which gives a reliable indication of the point of readiness.

Storage of Jam

The surface of jam is susceptible to mould growth; yeast cannot grow or thrive. There is risk of mold development and fermentation, with alteration in the taste. Permitted preservatives may be used at levels sufficient to inhibit the growth of yeast and moulds.

Jam should be stored in a fairly cool place or else some moisture evaporates resulting in surface graining and shrinkage of jam. It should be stored in a place having RH of ~ 80%.

Jelly

Jelly is semi solid products obtained by boiling a clear, strained fruit juice with sugar and acid to a thick consistency, jelly total soluble solids not less than 65% and acidity 05-0.7 percent.

For proper texture, jellied fruit products require the correct combination of fruit, pectin, acid and sugar. The fruit gives each spread its unique flavor and color. It also supplies the water to dissolve the other ingredients and furnishes some or all of the pectin and acid. Good-quality, flavorful fruits make the best jellied products.

Pectins are substances in fruits that form a gel if they are in the right combination with acid and sugar. All fruits contain some pectin. Apples, crabapples, gooseberries and some plums and grapes usually contain enough natural pectin to form a gel. Other fruits, such as strawberries, cherries and blueberries, contain little pectin and must be combined with other fruits high in pectin or with commercial pectin products to obtain gels. Because fully ripened fruit has less pectin, one-fourth of the fruit used in making jellies without added pectin should be underripe.

The proper level of acidity is critical to gel formation. If you use too little acid, the gel will not set. If you use too much acid, the gel will lose liquid (weep). For fruits low in acid, add lemon juice or other acidic ingredients as directed. Commercial pectin products contain acids that help ensure gelling.

As per FSSA, Fruit Jelly means the product prepared by boiling fruit juice or fruit(s) of sound quality, with or without water, expressing and straining the juice, adding nutritive sweeteners, and concentrating to such a consistency that gelatinization takes place on cooling. The product shall not be syrupy, sticky or gummy and shall be clear, sparkling and transparent.

The specific requirements are as shown below: Total soluble solids (m/m) - Min. 65 per cent

The product shall be manufactured from minimum 45 per cent, by weight, of original prepared fruit, exclusive of any added sugar or optional ingredients of finished product.

Fruits are classified as follows according to their pectin and acid contents:

- Rich in pectin and acid: Grape, sour guavas, sour black berries lemon, oranges (sour), plum (sour).
- Rich in pectin but low in acid: Apple (low acid varieties), unripe banana, sour cherry, pear, ripe guava.
- Low in pectin but rich in acid: Apricot (sour), sweet cherry, sour peach, pineapple and strawberry.
- Low in pectin and acid: Ripe apricot, peach (ripe), pomegranate, rasp berry, strawberry.

Low pectin containing fruits require pectin addition like apricot, pineapple, strawberry, raspberry, etc. For making jelly firm fruits are selected and cut into thin slices after washing. Thin cut slices are boiled with water which is 1.5 times the weight of fruit for about 20-30 minutes. During this process citric acid is added (2g per kg of fruit) which is followed by straining of extract.

Pectin test is performed with the prepared extract for knowing the amount of sugar which is to be added to initiate gel formation. Following two tests can be performed:

- **Alcohol test**: In this method pectin is precipitated with addition of alcohol. 3 teaspoonfuls of alcohol are poured gently down the side of the beaker containing one teaspoonful extract which is rotated for mixing and allowed to stand for a few minutes.

  1. If extract is rich in pectin, a single, transparent lump or clot will form. An equal amount of sugar is to be added to the extract for preparation of jelly.
  2. If extract contains a moderate amount of pectin, the clot will be less firm and fragmented. Three-fourths the amount of sugar is to be added.

3. If extract is poor in pectin, numerous small granular clots will be seen which might be because of more amount of water in extract. Further concentration of diluted extract can solve the above problem.

- **Jelmeter test**: The jelmeter is held in the left hand with the thumb and forefinger. The bottom of the jelmeter tube is closed with the little finger. The strained extract is poured into the jelmeter till it is filled to the brim. While still holding the jelmeter, the little finger is removed from the bottom end and the extract is allowed to drip for exactly one minute, at the end of which the finger is replaced. The reading of the level of extract in the jelmeter is noted. This figure indicates how many parts of sugar are to be added to one part of juice. After the pectin test has been completed required amount of sugar is added to extract and boiling of mixture is performed. This mixture is judged for end-point in the same manner as performed for jam after which remaining citric acid and color is added. Jelly is hot filled in sterilized bottles. Waxing and capping of bottle is done and kept for ambient storage.

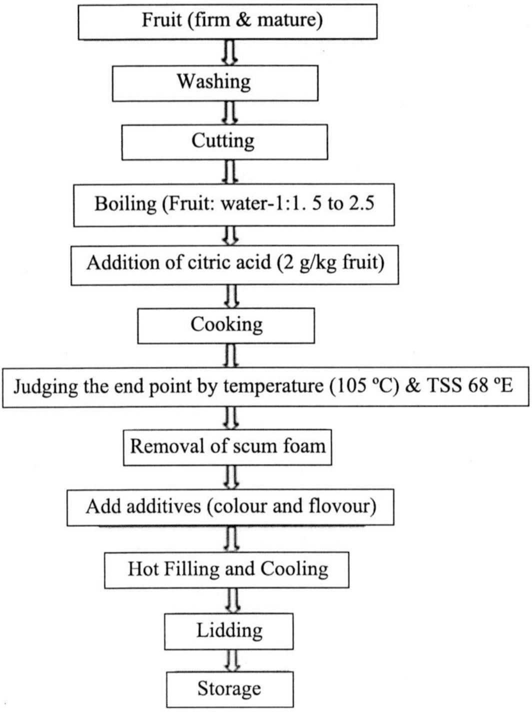

Figure 25: Flowchart of processing of jelly

Defects in Jelly

Crystals in Jelly: Sugar crystallization may result from:

- Too much sugar
- Too little acid
- Overcooking of jelly
- Too long a delay in sealing the container of jelly.

Crystals may sometimes be found in jelly because during the boiling, syrup spatters on the side of the pan and dries up. In subsequent pouring of the finished product, these crystals (act as seed material) are carried into the glasses of jelly.

Cloudy jelly: Usually occurs with the red juices. It is caused by imperfect straining. Restraining of juice without pressure brings a lower yield, but ensures a clear product.

- Use of non-clarified juice or extract.
- Use of immature fruits: Unripe, green fruits contain starch which is not soluble in the juice and therefore, gives it a cloudy appearance.
- Over-cooking: Due to high viscosity such jellies are gummy or sticky do not become clear after pouring into containers.
- Over-cooling: If the jelly is cooled too much, it becomes viscous and sometimes, lumpy and is always almost cloudy.
- Presence of scum: The jelly becomes cloudy when the scum is not removed.
- Faulty pouring: When jelly is poured into containers from a great height, some air gets trapped in the form of bubbles and makes the jelly opaque. Pouring vessel should not be held more than about 2.5 cm away from the top of the container.

Failure to Gel

Causative factors: Improper balance of pectin, acid, sugar and mineral salts, which may come about in several ways:

- Fruit used may lack sufficient pectin or acid or both
- Overcooking may destroy so much pectin that a gummy mass is formed
- Undercooking, due to insufficient concentration
- Too much water used for extraction of juice, so that the proportion of sugar is too great for the pectin and the long time required for evaporation may destroy some of the pectin.

Tough Jelly: The jelly becomes tough or stringy when too little sugar is used for the quantity of fruit juice used or when boiling is continued after the jellying point has been reached.

Syneresis of jelly: The phenomenon of spontaneous release of fluid from a gel is called syneresis and is caused by several factors:

- Excess of acid: Excess acid causes hydrolysis or decomposition of pectin. It can be prevented by mixing either some quantity of juice low in acid or more of pectin, so that a larger quantity of sugar can be added which helps in reducing the acidity and increasing the volume of jelly.
- Insufficient pectin: This results in the formation of a pectin network which is not sufficiently strong to hold the sugar syrup.
- Premature gelation: During the pouring of jelly into containers pectin network breakdowns and thus the jelly becomes weak and remains broken.

Marmalade

As per FSSA, Marmalade means a product prepared by boiling sound fruits with peel, pulp and juice, with or without water, added nutritive sweeteners and concentrating to such a consistency that gelatinization takes place on cooling of the product. It shall not be syrupy, sticky or gummy and shall be clear and transparent. Marmalade is a fruit jelly in which peels are suspended. Marmalades are prepared usually from citrus fruits. The protein content of marmalade is slightly higher. FPO specification for marmalade area same as mentioned for jelly.

Two types of marmalades can be prepared (i) jelly marmalade, and (ii) jam marmalade.

- Jelly marmalade: The following combinations give good quality of jelly marmalade:
  i) Sweet orange (Malta) and *khatta* or sour orange (*Citrus aurantium)* in the ratio of 2: 1 by weight with shreds of Malta orange peel
  ii) Mandarin orange and *khatta* in the ratio of 2: 1 by weight with shreds of Malta orange peel
  iii) Sweet orange (Malta) and *galgal* (*Citrus iimonia*) in the ratio of 2: 1 by weight with shreds of Malta orange peel

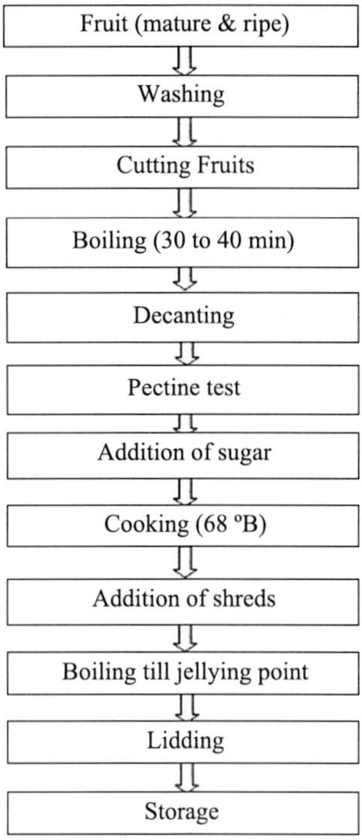

Figure 26: Flow chart for processing of marmalade

Jam marmalade: The method of preparation is practically the same as that for jelly marmalade. In this case the pectin extract of fruit is not clarified and the whole pulp is used. Sugar is added generally in the proportion of 1:1. The pulp-sugar mixture is cooked till the TSS content reaches 65 per cent. For marmalade making, fruit is peeled and peel is cut into fine shreds of 0.2cm width and 2-2.5 cm long. Fruit is cut into thin slices or convert into pulp according to final product required (jelly marmalade or jam marmalade). Slices or pulp is boiled 2-3 times its weight of water for 50-55 minutes. Extract is strained and pectin test is performed as done in jelly making. As the required pectin content is achieved, sugar is added and cooking of mixture at 103 to 105°C is performed. Prepared shreds which are boiled in different batches of water 3-4 times for making them soft and to remove bitterness are added at the rate of 62 g per kg of extract. Boiling is continued till the end- point is achieved which can be estimated by any of the methods discussed earlier in jam making. Product is finally cool to around 80-85°C and filled in sterilized bottles which are sealed and kept for storage.

Problems in marmalade making: Browning is one problem with marmalades during storage. It can be prevented by addition of small quantity of potassium metabisulphite. KMS dissolved in a small quantity of water is added to the marmalade while it is cooling. KMS also prevents spoilage due to moulds.

Chutney

Chutney' is a relish-type condiment; its increasing popularity reflects the inclusion of ethnic world cuisines in the Western diet.

The term 'chutney' includes several different varieties of sauce- type foods, drawn from traditional East Indian cuisine. The main ingredient may be an herb such as cilantro or mint; a flavoring ingredient such as coconut, onion, ginger, tamarind; or, in the most common form, chopped fruit or vegetables, simmered with spices, onion, sugar and vinegar. Fruit-based chutneys are usually cooked, then canned or refrigerated. Other chutneys like cilantro, onion, coconut, etc. are usually eaten fresh, with minimal, if any, cooking.

Fruit chutneys are most commonly available and varieties include mango, apple, apricot, cranberry, date, papaya, peach, pear, pineapple, plum, tomato and mixed fruit, to which raisins and nuts may be added to complement the texture. The result is a sweet- sour-spicy-hot versatile blend—an adventure for the taste buds.

According to FSSR (2011), Chutneys are the product prepared from washed, clean, sound raw fruits and/or vegetables of any suitable variety, which have been peeled, sliced or chopped or shredded and comminuted followed by cooking in vinegar, sweetened with sugar or dried fruit and flavoured with spices and simmered for a long time until they reach the consistency of jam. Chutneys should be smooth and pulpy with a mellow flavor and are best left to mature to blend flavours for at least one month. Chutney of good quality should be palatable and appetizing and improves the digestion. In general, chutneys fall into two distinct categories: freshly made preparations for immediate consumption, and cooked preparations intended to keep as long as a year, which can be grouped further according to their saltiness, sweetness, sourness, or spiciness. Many recipes combine several elements of these basic flavors, and textures range from coarsely chopped preserves to smooth sauces. Some of the most common chutneys are those made with mangoes, coconut, sesame, peanuts, or the ground leaves of herbs, especially mint or coriander. Mango chutney is an important food product exported from India to many countries. Apple and apricot chutney are also very popular in the country.

The product may contain permitted food additives and shall confirm to the microbiological requirements as per food safety and standards rules. It shall meet the following requirements:

Table 9: Total soluble solids

Total soluble solids	
Fruit chutney	Not less than 50%
Vegetable chutney	Not less than 25%
Hot and sour (spicy chutney)	Not less than 25 %
Fruits and vegetable content(m/m)	Not less than 40%
pH	Not more than 4.6
Total ash (m/m)	Not more than 5%
Ash insoluble in hydrochloricacid (m/m)	Not more than 0.5%

Procedure for Preparing Chutney

The fruit/vegetable is cut into slices of suitable size. Softened by dipping in boiling water. Slow cooking is preferred to yield better product than that of bristle heating at high temperature. Onion and garlic are added at the start to mellow their strong flavour. Spices are coarsely powdered and added. Spices can also be added by placing all ingredients in a cloth bags, loosely tied and placed in the mixture during cooking. Vinegar extract of spices can also be added. The vinegar is added just little before final stage of boiling. In place of vinegar, acetic acid can also be used as source of acidity. The product is cooked to a consistency of jam and filled hot into sterilized jars. The product can be pasteurized and processed at 82 C for 30 minute. The storage of chutney in done at ambient temperature in cool and dry place.

Sweet Mango Chutney: Mango chutneys are very special. The flesh softens and stays intact and blends easily with sugar, spices and vinegar without losing its flavour. Mango chutney is very dark in colour, rich, hot and fruity that still maintains the quality of mango.

Equipments: Iron and copper made utensils and equipment should not be used as they are acted upon by vinegar. Small amounts of these metals act as catalyst for rancidity in the product. Furthermore, these metals react with spices and tannins of the fruit resulting in the formation of black compounds which adversely affect the colour, flavour and taste of the product. Therefore, it is wise to use equipment made from nickel or stainless steel (SS) or strengthened glass such as Pyrex.

Stage in process: Essential	Stage in process: Optional	Quality Assurance	Equipment Required	Notes
Harvest fruit		Check for full maturity but not over-ripe		Picked carefully to reduce bruising etc. which would cause spoilage and loss of quality.
Wash			Wash tank	Use clean water to remove stones, leaves or soil
Sort/ grade		Essential to produce uniform quality products		Sort by hand to get similar colour or maturity.
	Peel	Check that all traces of peel are removed	Knives, peelers	Most fruits require peeling, but some chutneys include un-peeled fruits. By hand using sharp stainless steel knives or small peeling machines for some types of fruits.
	Cut/slice/core	Check for uniform sized pieces	Knives, slicers, dicers	Most fruits are cut into small strips. It is important to reduce the size of pieces to obtain uniform mixtures of similar sized fruits and/or vegetables. Cutting also allows faster penetration of sugar. This may be done by hand, but small scale equipment is available to increase production rates (Figure 9).
Mix	← Spices, sugar vinegar		Mixing bowl, scales	For each 500g of mango, spices such as ginger (0.5g), mustard (0.5g), chilli (1g), garlic (0.5g) and salt (1g) are washed, ground and mixed with vinegar (250ml) and sugar (300-500g). In some recipes, 7% oil is added.
	Heat	Check time and temperature	Heat source, boiling pan, (refractometer optional)	Boil until sufficiently thick and concentrated. Heat destroys enzymes and micro-organisms which together with high sugar level, spices & acid from the vinegar, helps to preserve product.
	Ferment	Check temperature and absence of insects	Covered pans	Natural fermentation over 2-3 days produces acids to assist preservation. Traditionally such products were not pasteurised.
Fill & Seal	← Jars/ lids	Check fill-weight and correctly sealed pack	Funnel or paste filler, scales	Packaging contains the contents and prevents re-contamination. Hot fill into pre-sterilised jars and hand seal lids - see text for alternatives.
Cool, Label & Store		Check label & storage conditions		In boxes, on racks, with a cool dry atmosphere.

Figure 27: Processing chart for making mango chutney

- Bottling: Strong glass jars fitted with plastic screw tops or plastic-coated metal screw tops are used. Chutneys should be covered with waxed paper discs before the screw tops are put on and make sure that no air bubble should trapped underneath by pressing well. All jars should be thoroughly cleaned and sterilized before use. Chutneys should be stored in a dry, cool, well-ventilated place for no longer than six months, afterwards vinegar may start to evaporate and the chutney will dry out.

- Chopping and Mincing: Chopping board and sharp knives are essential for vegetables and fruits. Dried fruits can be minced along with other ingredients using hand or an electric mincer.
- Cooking: Special preserving or saucepans made of stainless steel or aluminum are ideal. Saucepans should be large enough to accommodate all the ingredients to about two- thirds full. Large SS or wooden spoons are ideal for stirring the ingredients while cooking.

Methods for making chutney: chutneys are usually made with a mixture of fruits and vegetables.

- Initial preparation: The fruit or vegetables are thoroughly washed and cut in to slices or pieces of suitable sizes and softened by boiling in water. After this, they can be chopped by hand or by a mixer. Hard fruit are minced by mincer which is difficult to chop. Dried fruits may be softened by soaking in vinegar up to 12 hours in advance.
- Cooking: In simple chutney making procedure, all the ingredients are put together in the pan and brought up to boil and then gently simmered to form a thick puree with a consistency of jam. Sugar produces a darker colour when cook for a long time. For pale colour chutney, white sugar should be added after all the ingredients have cooked out. During initial cooking, chutney will be stirred in a span of every ten minutes, but towards the final stage, chutney become thicker thereby it requires frequent stirring to prevent it from sticking to the bottom of the pan. Chutneys are ready, when they are thick, bubbling and leave the impression of the dragged spoon which persists for a few seconds.
- Bottling: As soon as the chutney stops bubbling, it is ladled in to warm sterilized jar followed by pressing after each ladleful to dislodge the air bubble. Any drips or stickiness on the rim is wiped off and chutney is covered immediately.

4. Storing: Chutneys improves and mellow with keeping, so they are left unopened at least one month or more.

Several Different Factors Contribute to the 'Preserved' Nature of this Product:

1. The acidity (low pH) of the chutney prevents growth of several spoilage and pathogenic bacteria, molds and yeasts. This acidity is derived from the added vinegar and the natural acids of the fruit.
2. Cooking the mixture to concentrate it lowers available moisture that is needed for microbial growth. The cooking step also kills most microorganisms that may be present.
3. Processing the filled jars in a canner uses additional heat to kill spoilage organisms that might contaminate the product as jars are filled and to produce a vacuum seal for later storage. If the two-piece canning lid is applied correctly, air is driven out of the headspace while the jars are in the canner and a vacuum seal is formed upon cooling. For most chutneys, a boiling water canning process is adequate, but other foods may require a pressure process.
4. During storage in the sealed jar, oxygen and additional microbial contamination is kept from the product. Too much oxygen left in the jar will cause interactions with food components that lead to quality losses (for example, undesirable changes in color, texture, and flavor).

Pickles

Pickle is an edible product preserved in a solution of common salt and vinegar. It is one of the most ancient methods of preserving fruits and vegetables. Pickles are good appetizers and add to the palatability of meal. They stimulate the flow of gastric juice and thus help in digestion. Several kinds of pickles are sold in Indian market. Mango pickle ranks first. Pickles can also be prepared from fruits and vegetables like lemon, amla, onion, cauliflower, cabbage, beans, cucumber, bitter gourd, jackfruit, turnip etc. These are commonly made in homes as well as commercially prepared and exported. Fruits are generally preserved in sweetened and spiced vinegar, while vegetables in salt.

Pickling Process

The preservation of food in common salt or in vinegar is called pickling. Pickling may also be the result of fermentation by lactic acid forming bacteria, which are naturally present in large numbers on the surface of fresh vegetables and fruits. These bacteria can grow in acid medium and in the presence of 8-10% salt solution, whereas the growth of majority of undesirable organisms is inhibited. Lactic acid bacteria are most active at 30°C, so this temperature should be maintained, as far as possible, in the process of pickling. Pickling is done in two stages.

Stage I can be done by any of the three following methods:

i) Fermentation with dry salting,

ii) Fermentation in brine, or

iii) Salting without fermentation.

Stage II is finishing and packing.

Fermentation with Dry Salting

In this method, the vegetable is treated with dry salt. The salt extracts the juice from the vegetables and forms the brine, which is fermented by lactic acid bacteria. The method of dry salting in general is as follows: The vegetable is washed, drained, weighed for preparing pickles. Several alternate layers of the prepared vegetable and salt (20-30 g of dry salt/ kg vegetables) are kept in a vessel which is covered with a cloth and a wooden board and allowed to stand for 24 hrs. During this period brine is formed by osmosis. As soon as the brine is formed, the fermentation process starts and the CO_2 begins to evolve. When fermentation is over, gas formation stops. Under favourable conditions fermentation is completed in 8-10 days, however in cold weather it may take 2 to 4 weeks. When sufficient lactic acid has been formed,

lactic acid bacteria stop to grow and no further change takes place in vegetables. However, precaution should be taken against spoilage by aerobic microbes, because in the presence of air "pickles scum", a kind of wild yeast, is formed which brings about putrefaction and destroys the lactic acid. Therefore, the product may be preserved and kept by excluding air.

Fermentation in Brine

Steeping of the vegetables in a salt solution of pre-determined concentration for a certain length of time is called brining. When vegetables are placed in brine, it penetrates in the tissues of the vegetables and soluble material present in vegetable diffuses into the brine by osmosis. The soluble material includes fermentable sugars and minerals. The sugars serve as food for lactic acid bacteria, which convert them into lactic and other acids. The acid brine thus formed acts upon vegetable tissues to produce characteristic taste and aroma of pickle. The amount of brine required is usually half the volume of vegetables. Brining is the most important step in pickling. The growth of most spoilage organisms is inhibited by brine containing 15% salt. Lactic acid bacteria, which are salt-tolerant, can thrive in brine of 8-10% strength though fermentation takes place well even in 5 per cent brine. In a brine containing 10% salt, fermentation proceeds somewhat slowly. Fermentation takes place to some extent up to 15 % but stops at 20% brine strength. It is, therefore, advisable to place the vegetables in 10 % salt solution for vigorous lactic acid fermentation. After fermentation process, the salt content is now increased gradually, so that by the time pickle is ready, salt concentration reaches 15%.

Salting Without Fermentation

Vegetables are washed, prepared, and is mixed with salt (250 g/kg of prepared material). This high salt concentration will inhibit the fermentation. Vegetables packed with large amount of salt get cured. Then, they are drained and excess of salt is removed by soaking them in cold or warm water. Thereafter, the vegetables are stored in plain vinegar of 10% strength for several weeks. Vegetables can also be stored in sweetened and spiced vinegar. The spices can be added in the ground form or essential oil of spices may be added to impart the spice flavor. At present, pickles are prepared with salt, vinegar, oil or with a combination of above ingredients with spices. These methods are discussed below

Preservation with Salt

Salt improves the taste and flavour and hardens the tissue of vegetables and controls fermentation. Salt content of 15% or above prevents microbial spoilage. This method of preservation is generally used only for vegetables, which contains very little sugar. Since the sugar content is less, sufficient lactic acid cannot be formed by fermentation to act as preservative. However, some fruits viz., mango, lemon, etc. are also preserved with salt.

Preservation with Vinegar

This technology is based on the addition of food grade vinegar which has a bacteriostatic action in concentrations up to 4% acetic acid and bactericidal action in higher concentrations. Vegetables preserved in vinegar need to reach a final concentration of 2-3% acetic acid in order to assure their preservation. To achieve this final concentration, a 6-9% acetic acid vinegar is used, as related to the specific ratio of vinegar: vegetable. This higher concentration treatment helps to expel the gases present in the intercellular spaces of vegetable tissue. In vinegar pickles, salt (2-3%) and sometimes sugar (2-5%) are also added. If the vinegar concentration is lower than 2%, vinegar pickles need to be submitted to a pasteurization in order to assure their preservation. Mango, garlic, chilies, etc. are preserved as such in vinegar. Vinegar pickles are the most important pickles consumed in other countries.

Preservation with Oil

Oil pickles are highly popular in India. They are highly spiced. In India, mustard oil, rapeseed oil, sesame oil are generally used. The fruits or vegetables should be completely immersed in the edible oil. Cauliflower, lime, mango and turnip pickles are the most important oil pickles. The pickle remains in good condition for one to two years if handled properly.

Preservation with Salt, Vinegar, Oil and Spices

This method combines the advantages of fermentation action of salt and the preservation action of both vinegar and oil. The flavoring property of spices is also made use of. The spices are usually fried in oil and mixed to the prepared fruit/ vegetable before the addition of vinegar. The spices can be added separately or in the form of spice vinegar.

Common Defects Noticed in Different Pickles During Storage

- **Blackening**: It is mainly due to the presence of iron which enters through the brine or from the equipment. Blackening may also be caused by the action of microorganisms.
- **Dull and faded pickle**: It is due to either insufficient curing or use of water of inferior quality.
- **Shriveling:** It occurs when vegetables like cucumber are placed directly in a very strong solution of salt or vinegar. To avoid this, use weak solution at the start and increase gradually.
- **Softness and slipperiness**: It is the most common type of spoilage caused by the action of bacteria. Use of weak brine or improper covering with brine results in these defects. Thus, this defect can be checked by using brine of proper strength and keeping the pickle well below the surface of the brine.
- **Bitter Taste**: It is due to the use of very strong vinegar or cooking the spices for a long time or by using spices in excess.
- **Scum formation**: It is due to growth of film yeast on the brine kept for curing of vegetables. The scum may be thin or thick varying from imperceptible film to a thick wrinkled layer. It retards the formation of a lactic acid and helps in the growth of putrefactive bacteria which makes the vegetable soft or slippery. The scum may be removed as soon as it is formed. Use of 1% acetic acid to the brine prevents the growth of wild yeast on the brine, without hindering the formation of lactic acid.
- **Cloudiness:** In many fermented solid vegetables like onion, cucumber, olives etc the vinegar becomes cloudy and turbid, thus spoiling the appearance of the product. It is due to non-penetration of acetic acid from vinegar deep in to the tissues to check the activity of bacteria or other micro- organisms. This results in fermentation which make the vinegar cloudy. Use of brine and vinegar of proper strength prevents cloudiness.

- **Blemishes in pickles**: In onion pickle in brine, white blotch is seen under the first layer of the skin which is termed as blemish. Blemishes are generally caused by improper fermentation or non-removal of all brine prior to the final pickling of cured onion in vinegar.

FPO specifications for pickles: According to FPO specifications the minimum percentage of salt (w/v) in pickle in brine shall be 12% while for pickle in citrus juice the acidity as citric acid shall not be less than 1.2 per cent and only citrus fruit juices should be used. For oil pickles any edible vegetable oil like rapeseed, mustard, olive etc. can be used. Other general characteristics for these pickles include use of wholesome fruit and vegetables which are free of fungal or insect attack or any rotting. All ingredients shall be thoroughly clean and free from of extraneous matter. Only substances that may be added are spices, salt, sugar, jaggery, onions, garlic, benzoic acid, soluble calcium salts. Pickles shall be free from added copper, alum, mineral acids or other preservatives.

Sauces

Sauces and chutneys are usually made from the same raw materials, spices and flavours; however, difference is that, all sauces are sieved and as a result, are thinner and of smoother consistency than chutneys. The sieving is done to remove the skin, seeds and stalks of fruits, vegetables, and spices and to give a smooth consistency. Here the cooking process is longer compared to the chutney due the use of fine pulp or juice. Sauces are generally of two kinds, and they are the thin sauces and the thick sauces. A good sauce whether thin or thick, have a continuous flow with no skin, seeds and stalks of the fruit and/ or vegetables and spices used for its preparation, and possesses pleasant taste and aroma. The details of thin and thick sauces are given below:

Thin Sauces

Thin sauces mainly consist of vinegar extract of various flavoring materials like, spices and herbs. Their quality depends mostly on the piquancy of the material used. Some sauces are matured by storing them in wooden barrel or casks. During storage they develop flavour and aroma. Freshly prepared products have often a raw and strong taste and they should, therefore, be matured by storage.

Preparation: For the preparation of thin sauces of high quality, the spices, herbs, fruits, and vegetables are macerated in cold vinegar. Sometimes, they are also prepared by boiling them in vinegar. The sauce is filtered through a fine or coarse mesh sieve of non- corrodible metal, according to the quality desired. The skin, seeds and stalks of fruits, vegetables and spices used, should not be allowed to pass through the sieve as they spoil the appearance of the sauce. The usual commercial practice is to prepare vinegar extracts of each kind of spice and fruit separately, either by maceration or by boiling in vinegar and then blending these extracts suitably before filling the sauce into barrels for subsequent maturation.

Soya sauce made from soybeans and Worcestershire sauce made from tamarind are examples of thin sauces. The Worcestershire sauce is utilized in the preparation of cocktail also.

Thick Sauces

A sauce which does not flow freely and which is highly viscous is called a thick sauce. On the other hand, thin sauces are less viscous in consistency. Thick sauces also contain more of sugar and less of acid. Generally, spices and coloring added are practically like those in the case of thick ketchups and sauces. It should contain at least 3% acetic acid to ensure its keeping quality. The acidity should not however, exceed 3.4% as otherwise the sauce would taste sharp. The sugar content may usually be varied from 15-30% according to the kind of sauce made. Usually malt vinegar is used. In addition to contributing to acidity of the sauce, it also improves its flavor. The sweetness is derived partly from dates, raisin, apple, and tomato and partly from the sugar added. The colour of the sauces varies with the raw material use. Sometimes a little caramel is added.

Preparation: The manufacturing process is the same as for chutneys. Thickening agents are also added to prevent or retard sedimentation of solid particle in suspension in the sauce. In this country apple pulp is often used for this purpose. The starch obtained from maize, potato, arrow root, sago and rye are also used as thickening agents. Indian gum, gelatin, Irish moss, pectin and other similar substances can also be used subject to the food laws of our country. Tomato sauce and apple sauce are some of the examples for thick sauce.

The importance of quality and its considerations has been discussed in detailed in Unit 1. Now we will see the FPO specifications for the products discussed in the present unit. FPO specifications for tomato products, pickles, chutneys, and sauces are as follows:

Prevention of Black Neck

- Fill hot sauce at temperature not less than 85^0C.
- Leave less head space in bottles (more air in bottles will result in more blackening).

- Reduce the chances of iron contamination.
- Partial replacement of sugar by corn syrup or glucose may prevent blackening.
- Store bottles in horizontal or inverted position to diffuse the entrapped air/oxygen.

Fruit Jeather

Fruit leather is made by drying thin layers of pureed fruit in the oven or dehydrator. Sometimes called fruit rolls or taffies, fruit leathers make delicious, wholesome and nutritious high-energy snacks for backpackers, campers and active children. They are relatively light in weight, easy to prepare and a good way to use left-over canned fruit and slightly over-ripe fresh fruit. Fruit leathers can be eaten as is, or made into a beverage by combining 5 parts water with 1 part leather in a food blender. They also can be used in pie fillings, in cooking and as a dessert topping. Most fruit or combinations of fruits can be used. Apricots, apples, grapes, berries, bananas, pineapples, oranges, pears, peaches, plums, melons, and most tropical fruits can be blended and dried to make fruit leathers. Grapefruit and lemons are not recommended because they turn bitter when dried. Drying is one of the least exact ways to preserve foods. The length of drying time will depend on the equipment used and the humidity of the air. In the past, recommendations for preparing fruit leather from both fresh and cooked fruit have been given. However, because of increasing concerns with bacteria such as *Escherichia coli O157:H7* (*E. coli O157:H7*) being able to survive the drying process if present, it's best to heat the fruit to 160^{0}F before drying. Preheating also stops the maturing action of enzymes in the fruit, helps preserve the fruit's natural color and speeds the drying process.

Leather Preparation

Select ripe or slightly over-ripe fruit. Sort and thoroughly rinse or scrub the fruit under running water. Remove and discard blemishes or defective parts. Peel tough-skinned fruits such as winter apples, oranges, peaches, pears and tomatoes, if desired. Pit and core fruit as needed. Remove seeds from grapes.

Hull Strawberries

Cut fruit into chunks and place in the top of a double boiler. Place water in the bottom of the double boiler and bring to a boil. Cover and steam for 15 to 20 minutes or until the fruit is soft and a thermometer placed in the fruit mixture registers at 160 ^{0}F. The fruit mixture may also be cooked in a microwave oven. Place cut fruit in a glass casserole. Cover and microwave on full power (high) for 6 to 8 minutes per two cups of fruit, stirring every 2 minutes. Place cooked fruit in blender. Add ½ teaspoon of ascorbic acid crystals or 2 tablespoons lemon juice per 2 cups of fruit to protect the color and help destroy bacteria during drying. If desired, add 1 to 2 tablespoons of sugar, corn syrup or honey per 2 cups of fruit. A small amount of spice (¼ teaspoon cinnamon or a dash of nutmeg) may also be added per 2 cups puree, for taste variety.

Shortcut canned method: Substitute canned fruit or strained baby food without tapioca for the cooked fruit above. Canned applesauce and strained baby fruit will not need to be pureed. Other canned fruits will need to be drained and pureed in a blender, food grinder or by hand. Canned fruits are already processed, which destroys bacteria and stops enzyme action. Thus, the addition of ascorbic acid or lemon juice is not necessary. Canned fruits, such as applesauce, can be mixed with more expensive fresh fruits to help stretch the fruit concentrate and soften the flavor of sharp-tasting fruits, such as cranberries. The addition of applesauce to juicy fruits also eases drying.

Leather Drying

Spray a cookie sheet or similar flat tray with vegetable spray or line with plastic wrap. Make sure the cookie sheet or tray has an edge to prevent spillage of the puree. Spread the fruit concentrate evenly over the pan surface to a depth of 1/8 to 1/4-inch. Two cups of puree is enough to cover a 12- by 17-inch cookie sheet. Dry the concentrate in an oven or dehydrator.

- **Oven Drying**. Set oven at the lowest setting (140^{0}F to 145^{0}F). Place the trays of puree on the oven rack and leave the door open 2 to 6 inches, depending on the oven door. Check the oven temperature periodically with a thermometer to be sure the air temperature of the oven is at the desired level. If necessary, turn off the oven for a short time to reduce the temperature. The fruit concentrate should dry in 4 to 10 hours. Test frequently for dryness.
- **Dehydrator drying**. Place sheets or trays of fruit concentrate in the dehydrator. Set temperature control at 140^{0}F to 145^{0}F or follow manufacturer's directions. Test frequently for dryness. Drying time will be 4 to 10 hours. Test for dryness. Properly dried fruit leather will be translucent and slightly tacky to the touch, but easily peeled from the pan or plastic wrap. Test for dryness by touching the leather in several places; no indentations should be evident. Lift the edge of the leather, which will adhere tightly to the surface, and peel it back about an inch. If it peels readily, it is properly dried. If the leather has cooled, it may need to be warmed in an oven at 150^{0}F for a few minutes to help it peel away more easily. If the leather cracks or chips, it has dried for too long, but is still edible.

- **Storage**. After loosening the edge of the leather from the plastic wrap or pan, loosely roll the leather in plastic wrap or waxed paper in one piece. Store the roll in one piece or cut into 1-inch strips. Place the strips or rolls of leather in a plastic bag, glass container, paper bag or other container. Until the leather is completely dry, the container lid should not be tightened nor the bag opening twisted tightly. If the leather has not dried completely, it may become sticky or develop mold growth during airtight storage. Store fruit leather in a cool, dry, dark place. It will retain good quality for up to one year in the freezer, several months in the refrigerator, or one to two months at room temperature (70^0F).

Nutritional food values become concentrated in dried fruit, and so do calories. Since moisture is gone, the residue is concentrated. A 1- by 17-inch strip of applesauce leather provides approximately 40 calories, assuming 2 cups of canned sweetened applesauce were dried on a 12- by 17-inch pan.

Tomato Puree and Concentrate

Tomato pulp without skin or seeds, with or without added salt, and containing not less than 9 per cent of salt free tomato solids is known as "Medium tomato puree". It can be concentrated further to 'heavy tomato puree', which contains not less than 12% solids.

Preparation of tomato puree: Manufacturing steps fall into three successive categories: (i) obtaining tomato juice from the raw materials; (ii) juice concentration and (iii) tomato puree pasteurization. Tomato pulp and juice extraction are done from ripe tomatoes in the same manner as tomato juice preparation. Concentration of the pulp is carried out either in an open cooker or in a vacuum pan. In open cooking, most of the vitamins are destroyed and the product becomes brown. On the other hand, use of vacuum pans, nutritive value is preserved and browning is also reduced. However, this method is quite expensive. Ordinarily tomato juice can be concentrated to lower range in an open cooker, but for obtaining higher concentrations a vacuum pan is required. The end point is judged by determining the TSS. This is done by using either a specific gravity hydrometer or by a refractometer. Tomato puree pasteurization assures the microbiological stability of the product. For this purpose, the puree coming out from concentration equipment is passed continuously and in a "forced" mode through a tubular pasteurizer from which it emerges, at a temperature of 90-92^0C. The pasteurized puree is then filled hot in to cans or in glass receptacles.

This pulp-based concentrate can be concentrated to different levels of total solids to give a range of products:

- Puree 10^0 Brix
- Simple 16^0 Brix
- Double 29^0 Brix
- Triple $30\text{-}32^0$ Brix

The double and triple concentrates are prepared using a vacuum evaporator

1. Sort the fruit and select ripe tomatoes. Discard any over- ripe or mouldy fruits. Save the under-ripe fruits for a later batch.
2. Wash in clean water and drain. Cut the tomatoes into quarters, discarding any that are rotten inside.
3. Place the tomatoes in a large pot and cook over a medium heat, stirring with a wooden spoon occasionally. Add salt to taste if desired.
4. Continue heating until the total solids measure $6.5\text{-}6.8^0$ Brix (using a refractometer). Remove from the heat and allow to cool slightly.
5. Extract the juice by passing the sauce through a pulper. Pass the skins and seeds through the pulper a second time to remove as much juice as possible.

To make the puree

- Place the sauce back on the heat and concentrate until it is 10° Brix. Stir occasionally with a wooden spoon to prevent burning.
- Add 1 per cent salt, continue heating until it has dissolved, then remove from the heat.
- Fill clean, sterile glass jars to the top with the hot puree and cover with clean lids.
- Place the glass containers into a water bath that is at the same temperature as the bottles. The water must cover the jars. Sterilize in boiling water for 45 minutes from the time that the water starts to boil.
- Remove the water bath from the heat. Gradually add cold water to slowly cool the water bath. Or, leave the jars to cool in the water bath until the following day.
- Dry the jars, add labels and store in a cool dry place away from light. The product should be stable for at least 12 months.

To make the simple concentrate proceed as follows

- Concentrate the sauce until it reaches 16° Brix.
- Add 2 per cent salt, dissolve and remove from the heat.

- Fill clean, sterile glass jars to the top with the hot puree and cover with clean lids.
- Place the glass containers into a water bath that is at the same temperature as the bottles. The water must cover the jars. Sterilise in boiling water for 45 minutes from the time that the water starts to boil.
- Remove the water bath from the heat. Gradually add cold water to slowly cool the water bath. Or, leave the jars to cool in the water bath until the following day.
- Dry the jars, add labels and store in a cool dry place away from light. The product should be stable for at least 12 months.

Tomato Paste

The product with highest production volumes among concentrated products is tomato paste, which is manufactured in a various range of concentrations, with a minimum of 25 per cent and up to 44 per cent refractometric extract. The product is very similar to tomato puree except that the solid concentration is more. Tomato paste is the product obtained by removal of peel and seeds from tomatoes, followed by concentration of juice by evaporation under vacuum. Good quality tomato paste is a homogenous mass, with a high density, without foreign bodies (seeds, peel, etc.), with a red colour, and an agreeable taste and smell, close to those of fresh tomatoes. Preparation procedure is also similar to that of tomato puree. Here the tomato juice obtained is further concentrated so that it contains not less than 25 per cent tomato solids. This is known as tomato paste. On further concentration to 33 per cent or more of solids, it is called concentrated tomato paste. Paste of good quality must have a volatile acidity of maximum 0.15 per cent as lactic acid. An 8 per cent salt addition is accepted. The end point is judged by using a refractometer. The product with required TSS is then pasteurized and filled into receptacles in the same way as explained under tomato puree.

This pulp-based concentrate can be concentrated to different levels of total solids to give a range of products: This is a thick, bright red paste that is used to prepare many traditional sauces and soups. It can be sold from bulk containers, in jars, polythene bags of cans and has a shelf life of several months.

1. Sort the fruit and remove any under-ripe, moldy, or rotten tomatoes.
2. Wash in 8 liters of clean water per 10kg tomatoes.
3. Blanch in a water bath at 80^{0}C for 5 minutes. The tomatoes are blanched to soften the skins for peeling and to destroy enzymes and micro-organisms.

4. Pass through a pulper-finisher fitted with a 1mm sieve. This removes the skins and seeds from the fruit.
5. Measure the acidity of the pulp. If the pH is 4.0 or below, no lemon juice is needed. If the pH is above 4.0, lemon juice should be added until the pH falls below 4.0.
6. Add 0.3g sodium benzoate per 10 kg pulp to preserve the product after the bottle has been opened.
7. Concentrate the pulp by one of two methods:
   - Traditionally this is done by heating slowly in an open pan, stirring constantly to prevent burning. If this is done carefully, the bright red colour can be retained. Heating is continued until a paste with 30 per cent total solids is obtained. This is measured by a refractometer.
   - An improved method of concentrating is to place the pulp in a white cotton sack that is hung up for one hour to allow the water to drain out. The pulp loses half its weight. Add 2.5 per cent salt by weight of concentrate and mix thoroughly. Re-hang the sack for one hour until the pulp is one third of its original weight (until it is 30% total solids).
8. Fill the paste into bottles or cans, seal the containers.
9. Pasteurize the bottles at 90^{0}C for 45 minutes. Cool to room temperature.
10. Dry the bottles, label and store in a cool dry place away from direct sunlight.

Tomato Ketchup

Tomato ketchup is a popular condiment all over the world. Tomato ketchup is like tomato sauce except that it is thick in consistency. The number of spices added in case of tomato ketchup is considerably higher than in tomato sauce. Thick sauces made from fruits and vegetables other than tomatoes are not called ketchups.

Preparation: Clean, wholesome tomatoes of intense red colour and of meaty, not watery texture are used for sauce making. High acidity and a rich tomato flavour are additional desirable qualities. Sound tomatoes are washed very thoroughly, cored, sliced, heat crushed and pulped (through a pulper or juice extractor) to remove seeds and skins. Tomato pulp or paste is then cooked with the requisite quantities of spices, onions, garlic, sugar, salt and vinegar. The whole mass is concentrated to the required TSS (28%). A ketchup with a 28-30 per cent TSS has a better flavour than those with more than 30 per cent TSS. However, the higher TSS ketchup generally keep for a long time, once the bottle is opened. The end point is judged either by means of a specific gravity hydrometer or by a refractometer. In commercial practice, the juice is concentrated to one third of its original volume as determined with a gauge stick. Bottling: The mass is finally passed through the finisher, fitted with a very fine sieve, to remove any tomato fiber, and other impurities. The sieved ketchup is filled into clean, dry bottles. The ketchup should be filled hot (88^0C) to prevent browning and loss of vitamins during subsequent storage. Bottles are cleaned, labelled, and packaged for marketing. Preservation is assured either by use of preservatives or by pasteurization.

Fruit Cheese

Fruit butters and fruit cheese are old-fashioned names for preserves made with fruit puree and sugar. Generally, the fruits used for jellies are also the best ones for butter and cheeses and they can even be made from the pulp left in the jelly bag after the juice has dripped. The same equipment is required as for jam marking, with the addition of a fine nylon or plastic sieve.

Fruit butters are soft and spreadable. They are a particularly useful way for dealing with a glut of wild fruits and, as less sugar is required for this sort of preserve, they are very economical. Due to their lower sugar content, they generally have a storage life of only a few weeks and, once opened, should be eaten with a few days.

Fruit cheeses are made from a stiff fruit purée. This is a good way of preserving fruit which has a lot of pips or stones. They use a larger proportion of sugar to purée than fruit butters.

Fruit cheeses can be turned out in one piece and cut with a knife in place of cheese or, cut into small pieces, as a sweetmeat. They are also useful as a condiment to cold meat or poultry. Cheeses can be stored for up to 4 months and are often better if left to mature for 2 months before using

Making Fruit Cheeses and Butters

- Clean and roughly chop large fruits, discarding any parts that are bruised or diseased. Small fruits may be used whole.
- Place the fruit into the preserving pan and just cover with cold water. Simmer until soft, adding lemon juice or citric acid to fruits low in acid.
- Rub the softened fruit through a nylon or plastic sieve to produce a fine pulp. Weigh the pulp and pour it into a clean preserving pan or saucepan.
- To make a cheese add a similar weight of sugar to the pulp and stir until dissolved. Simmer for 1 hour or until the cheese thickens, stirring the mixture consistently.

- To make a fruit butter boil the pulp until it is thick, then stir in half the pulp's weight of sugar, plus spices if required. Simmer until the mixture is thick and creamy.
- Fruit cheese is ready when a spoon drawn across the pan leaves a clean line. Fruit butter is ready when no free liquid is visible and the surface is creamy.
- Pour cheeses into clean, hot, dry sterilized small wide- necked jars or moulds that have been greased with glycerine. Pour fruit butters into clean, hot, dry sterilized jars as when making jam.
- Wipe the jars or moulds clean and cover immediately whilst still hot.
- Leave to cool and label with contents and date made.

Dehydrated Wafers

Increasing the temperature of food makes its moisture evaporate, and air moving over the food carries the moisture away. A balance of temperature and humidity is needed to successfully dry foods.

Food dehydrators—either commercially made or homemade— give a good-quality dried product. Oven drying works well if you can set your oven to a temperature of 140 (60^{0}C) to 150^{0}F (65^{0}C). Open the oven door 2 (5 cm) to 3 inches (7.5 cm) to allow moisture to escape. A convection oven works well because it combines low heat with a fan to move the air. Room drying at room temperature works only if heat, humidity, and air movement are adequate.

Fruit flakes made by applying a moist mass of fruit pulp, sugar and starchy material to a hot smooth substrate in a thin layer, at least partially dehydrating said layer to form a cohesive film, peeling said film from said substrate and forming said film into an irregular structure, drying said film until it is brittle, and dividing said film into flakes.

Procedures

- Start with clean countertops and utensils. Wash hands with soap and warm water.
- All produce should be properly washed before it is consumed or preserved by rinsing, gently rubbing, or scrubbing with a clean vegetable brush under cold running water. Do not soak produce in water.
- Peeling is optional; however, the skin tends to toughen on apples and pears. The skin reduces surface area, preventing moisture from escaping. Sliced pieces will dry more quickly than fruit or vegetables left whole or cut in half. Pieces of the same size, shape, and thickness will dry evenly. Some thinly sliced fruits and vegetables, such as apple chips or zucchini

chips, will dry crisp. Pretreat foods as described below. Place pieces on drying racks without allowing them to touch or overlap. Place trays in a preheated dehydrator. Initially, the temperature can be set at 145^{0}F (62^{0}C) when there is surface moisture on the fruit or vegetable. After one hour reduce the temperature to 135 (57^{0}C) to 140^{0}F (60^{0}C) to finish drying. If the food is dried at a temperature that is too high, the outer surface will harden, preventing moisture from escaping from the center of the slice—this is called case hardening. Food shrinks when it is dried, so use a fine mesh for smaller fruits and vegetables.

- Pretreating Fruits: Some foods such as apples, pears, peaches, and apricots dry better when pretreated. Pretreatment reduces oxidation, giving a better color, reducing vitamin loss, and lengthening shelf life. Research studies have shown that pretreating with an acidic solution enhances the destruction of potentially harmful bacteria during drying. Place cut fruits in a solution of 3¾ teaspoons of powdered ascorbic acid (or crush 20 500-milligram vitamin C tablets) or ½ teaspoon of powdered citric acid in 2 cups of water for 10 minutes before placing on trays to dry. Equal parts of bottled lemon juice and water can be substituted for the above pretreatment.
- Other methods of pre-treating fruit include syrup blanching, water blanching, and sulfiting. Syrup blanching involves simmering the prepared fruit for 10 minutes in a syrup of 1 cup sugar, 1 cup white corn syrup, and 2 cups water and letting it stand in the hot syrup for 30 minutes before draining, rinsing, and placing on drying trays. Syrup-blanched fruit is sweeter but also stickier than fruit treated by other methods. Refer to a book on food drying for specific times and directions if you choose to blanch fruits. Although sulphites have been used in the past to prevent oxidation, this fact sheet does not include this method because sulphites are not recommended for use by individuals on restricted-sodium diets or who have asthmatic or respiratory conditions. Sulphated food should be dried outdoors for safety reasons.
- Some fruits such as blueberries and cranberries need to be dipped into boiling water to crack the skins. Be careful not to leave the fruit in the boiling water for too long or the fruit will turn to mush. Chill quickly after cracking skins and blot dry. Home-dried cranberries and dried blueberries are not like the commercially dried product, which is sweeter.
- Conditioning and Storing Fruits: Fruits are dry when they are pliable and no beads of moisture form when pressed between your fingers.

Condition dried fruit by packing it loosely into an air-tight glass or plastic container for several days to distribute the remaining moisture evenly. If condensation forms inside the container, further dehydration is needed.

- Unpeeled or uncovered fruits need to be treated to destroy insect eggs that might have gotten on the fruit. Heat dried fruit in the oven at 160^0F (71^0C) for 30 minutes or chill in the freezer at 0^0F or below for 48 hours. The shelf life of dried fruit is increased when it is stored in the freezer or refrigerator.
- Blanching Vegetables: Blanching in a solution that contains 1/4 teaspoon of citric acid per quart of water is recommended for most vegetables. Steam blanching is an option. This enhances the destruction of potentially harmful microorganisms and slows the enzyme reactions that will continue during drying and storage. Blanching also softens the cell structure, allowing moisture to escape, and allows the pieces to dry faster and later rehydrate faster. Blanched vegetables should be drained and placed on dryer trays. The heat from blanching will give them a head start in the drying process. Onions, garlic, peppers, and herbs do not need blanching.
- Testing Dryness and Storage: Vegetables are tough, brittle, or crunchy when dry and do not need conditioning. Store dried vegetables in air-tight containers to prevent food from absorbing the moisture in the air. Storing them in a dark place retains the vitamin content of the food.

Vegetable *Papads*

Potato *Papad*

Ingredients

Potato - 1 kgs. (medium sized)

Salt - less than 1 tea spoon (according to the taste) Red Chili Powder - less than a 1/2 tea spoon.

Oil - 2 table spoon.

Method

Wash and boil the Potatoes in the cooker in 2 cups of water for up- to 1 whistle. After the first whistle, boil the Potatoes for 3-4 minutes on a medium flame until they get soft or tender. Once the Potatoes are boiled and cools down, peel and grate them. Add salt, red chilli powder into the grated Potatoes and mash them. Now grease your hands and make small balls from the mixture. Make as much balls as you can from the mixture, but remember that all the balls should be of same size. Make all the balls and keep on placing them on a plate. You will be able to make 20-22 balls from the mixture of 1 kg potatoes. To roll out the Potato Balls you need 1x2 ft. thick polythene sheet and a big flat polythene -sheet is required to dry them under the Sun. Spread the big polythene - sheet on top of a bedsheet on the terrace floor. Put any heavy substance like bricks on all the four corners of the sheet so it does not misplace. To roll out the Papad take any thick transparent polythene sheet. Place the sheet on a rolling board in a way that half of it should be fall outside the board and half should be covering the inside of the board. You can even take two equal parts of a polythene - sheet. Place the potato ball on one sheet and the cover it with the other sheet. Apply some oil on the first polythene – sheet and on both sides of the potato ball. Place the first sheet on the rolling board and keep the greased potato ball on it. Now cover the potato ball from other sheet and press it with hands to flatten it. Roll the potato ball into thin flat disc as we do while making chapatis. Moreover, you can also press the potato ball with your fingers to give it a round shape.

Drying

Strip out the one side of polythene - sheet from the rolled out *papad* and put it over the big polythene - sheet from the side of the *papad*. Now press the *papad* with hands to stick it on the big polythene - sheet and strip out the other part of the polythene – sheet also. Press the papad against the big polythene sheet to settle it down. Place all the other *papad* on the sheet in the same manner.

Let the *papad* dry in the Sun. After 3-4 hours turn the sides of the *papad*, when they are little moist. If the *papad* will get completely dry from one side, then they might break on turning or get stick to the polythene - sheet. Usually, it takes a day for a *papad* to get completely dry, but if they are still little moist then collect them at the end of the first day and spread them again the next day to get completely dried.

Fruit Toffees

Fruit toffees are made by mixing fruit pulp with other ingredients like glucose, milk powder and edible fat. The fruit pulp is first concentrated to half its volume in a steam jacketed kettle or pan over steam bath. Generally, for one kilo gram of concentrated pulp, 160 g of glucose, 320 g of milk powder and 200 g of edible fat is added. This mixture is further heated to a thick consistency (75- 80^0Brix) followed by spreading it as a sheet of one cm thickness on a fat smeared flat tray and allowed to cool. Then these are cut into pieces (called as toffees) of desired size, wrapped and stored it in cool dry place.

According to FSSAI Fruit Bar/ Toffee means the product prepared by blending Pulp/Puree from sound ripe fruit, fresh or previously preserved, nutritive sweeteners, butter or other edible vegetable fat or milk solids and other ingredients appropriate to the product & dehydrated to form sheet which can be cut to desired shape or size.

Product shall comply with the following requirements: -

i) Moisture (m/m) Not more than 20 per cent

ii) Total soluble solids (m/m) Not less than 75 per cent

iii) Fruit content (m/m) Not less than 25 per cent

Soup Powders

Soup Powders means the products obtained by mechanical dehydration of fresh vegetables/ fruits juice/ pulp/puree of sound, vegetables /fruits and or earlier concentrated, dehydrated, frozen or processed fruits & vegetables, singly or in combination by blending with salt, nutritive sweeteners, spices and condiments and any other ingredients suitable to the product, as appropriate to the product and packed suitably to prevent spoilage.

2. The product shall comply with the following requirements

i) Moisture (m/m) Not more than 5 per cent

ii) Total soluble solids (m/m) (on dilution on ready to serve basis) Not less than 5 per cent

Soup powders are commonly prepared using the batch process, involving the mixing of dry powder ingredients such as vegetables, spices, flavors and yeast extracts, to which either oil or fat powder is added. Another method used to prepare soup powder is the spray drying of blended cooked and milled ingredients. Taking mushroom soup powder as a specific example, the follow preparation process is employed: a cooked blend of mushroom and concentrated cheese whey is mixed with fried spices (onion, garlic, ginger), hydrogenated fat and corn flour to provide the required thickening properties. The blend is then spray dried at an inlet temperature of 160–180 ^{0}C.47

Ungelatinized starch (5–35%) may be added to soup powder recipes to improve the thickness of the soup. The starch must be not only ungelatinized, but also tasteless, highly water-soluble, and in powder form with a particle size of < 74 μm. This is to ensure that the starch can gelatinize during cooking to provide the desired creamy properties of the final product. These properties should not be dependent on the level of solids in the soup ingredients. The incorporation of fatty substances is also important so that the hydration of starch can be delayed, allowing flavor to be imparted upon mixing with hot water.

3

Production of Pectin and Vinegar

Pectin

Pectin is a component of nearly all fruits and vegetables and can be extracted and used in food processing to form the characteristic gel in jams and marmalade. The richest sources of pectin are the peels of citrus fruits such as lime, lemon, orange and passion fruit or the pulp of apple (known as 'pomace' after juice has been extracted). Commercially, pectin is available as either a light brown powder or as a dark liquid concentrate. It is stable if stored in cool, dry place and it will lose only about 2 per cent of its gelling power per year.

Stage in process		Quality Assurance	Equipment Required	Notes
Essential	Optional			
Fruit		No special checks, but mouldy fruit should be discarded		Under-ripe fruit has a greater amount of pectin than over-ripe fruit.
Wash			Wash tank.	To remove pesticide residues, leaves, etc.
	Peel		Knives	Remove citrus peels and use for extraction. Other fruits are not peeled.
Cut		Check that pieces are small and uniform.	Knives, slicers	Peels are cut into thin slices (e.g. 1cm wide) and other fruits are coarsely chopped.
Mix	Water			Add water - approximately 8 times the weight of fruit.
Heat		Check the temperature and time of heating.	Boiling pan, heat source.	Heat gently to simmering and hold at that temperature for 20-30 minutes.
	Concentrate			Longer heating removes more water and concentrates the pectin so that it can be stored for a few days. However, over-long heating degrades the pectin and it loses some of its gelling power.
use in jam making process				

Figure 28: Process chart for pectin extraction

There are two main types of pectin:

- High methoxyl (HM) pectin that form gels in high solids jams (above 55% solids) in a pH range of 2.0-3.5; and
- Low methoxyl (LM) pectin, which do not need sugar or acid to form a gel, but instead use calcium salts.

These form a gel with a wide range of solids (10-80%) within broader pH range of 2.5-6.5. They are used mainly for spreads or for gelling agents in milk products. However, there are many different types of pectin within each group, such as 'rapid set' and 'slow set' pectin, that are made for different applications and it is necessary to specify carefully the type required when ordering from a supplier.

It is difficult to make specific grades of pectin or the powdered products at a small scale. Where attempts have been made to set up small commercial pectin units, the process has not been financially viable due to the expense of recovering solvents that are used to precipitate the pectin before drying. However, it is possible for jam-makers to increase the amount of crude (unrefined) pectin in their products by extracting it with water without using solvents. This is useful if the producer is making jam from fruits such as melon that are naturally low in pectin. When using pectin to make set preserves, such as jams, jellies etc., it is important that the pH is within the range of 3.0-3.5 and that they are boiled to at least 68 per cent solids content. Powdered pectin is added to fruit pulp at 3-6g per kg of final product, but it should first be mixed with about five times its weight of sugar to prevent lumps forming when it is added to the pulp or juice.

Vinegar

Vinegar is perhaps the oldest known product or fermentation. The word is derived from French 'vinaigre' meaning sour wine (vin = wine, aigre= sour). Vinegar is a liquid obtained by alcoholic and acetic fermentation of suitable materials containing sugar and starch (at least 10% fermentable sugar). It contains about 5 per cent acetic acid and has germicidal and antiseptic properties. In the trade, vinegar is labelled according to the material used in its manufacture, e.g., malt vinegar (from malt) and cider vinegar (from apple juice). The amount of acid in vinegar is expressed as 'grain strength' which is ten times the percentage of the acetic acid present in it, e.g., vinegar having 5 per cent acetic acid is termed as vinegar of '50 grain strength'.

Types of Vinegar

Vinegars are of two types

A. Brewed vinegars: Brewed vinegars are made from various fruits, starchy materials (potato) and sugar containing substances (molasses, honey) by alcoholic and subsequent acetic fermentation.

1. *Fruit vinegar*: Generally, apple, grape, orange, jamun, peach, pear, pineapple, apricot and banana are used. Vinegar made from apple juice is known as cider or apple cider vinegar, while that from grapes as wine or grape vinegar.
2. *Potato vinegar*: In this case starch is extracted from potato and hydrolyzed by the enzyme diastase before fermentation.
3. *Malt vinegar*: Malt vinegar is derived wholly from malted barley, with or without the addition of the cereal grain, malted or otherwise, the starch of which is saccharified by the diastase of the malt before fermentation. Distilled malt vinegar is prepared by distilling the malt vinegar. The product merely contains the volatile constituents of the vinegar from which it is derived. It is colourless and is generally used in the manufacture of pickled onions.
4. *Molasses vinegar*: In this case molasses is diluted to 16 per cent total soluble solids, neutralized with citric acid and then fermented.
5. *Honey vinegar*: It is prepared from low grade honey.
6. *Spirit vinegar*: Spirit vinegar is the product prepared by acetous fermentation of a distilled alcoholic fluid which in turn is produced by fermentation. It is usually made by alcoholic fermentation of molasses and then distilled prior to acetic fermentation. It is also called as grain vinegar. distilled vinegar, white vinegar or alcohol vinegar.
7. *Spiced vinegar*: Spiced vinegars are prepared by steeping the leaves or spices in an ordinary vinegar.
8. *Artificial vinegars*: Artificial vinegars are prepared by diluting synthetic acetic acid or glacial acetic acid to a legal standard of 4 per cent and are coloured with caramel. Artificial vinegars are also called as synthetic vinegar or non- brewed vinegar.

Steps Involved in Vinegar Production

Two distinct steps are involved in its preparation.

i) Conversion of the sugar in fruits, etc., into alcohol by yeast (alcoholic fermentation): The most efficient yeasts for fermentation of sugary substances and fruit juices into alcohol are *Saccharomyces ellipsoideus,*

S. malei and S. cerevisiae. For starchy substances. *S. cerevisiae* is the best. To obtain quality vinegar, it is essential to first destroy wild (naturally occurring) yeasts and other microorganisms by pasteurization, and then to inoculate pure yeast. The nutrients for the growth of yeast such as phosphates, ammonium and potassium salts and sugars are naturally present in fruit juices and in honey and molasses. The most favourable temperature for the growth of yeast is 25-27^{0}C. Fermentation becomes abnormal at 38^{0}C and ceases altogether at 41^{0}C and below 7.0

$$C_2H_5OH + O_2 \rightarrow CH_3COOH + H_2O \quad H = -493kj$$

ii) Conversion of alcohol into vinegar by acetic acid bacteria (acetification) Acetic acid fermentation is brought about by acetic acid bacteria (*Acetobacter spp.*) which are strongly aerobic but whose activity is greatly reduced by light. Acetic acid fermentation should, therefore, be carried out in the dark. The nutrients required for bacterial growth are generally present in the alcoholic liquor itself, but in the case or distilled alcohol, malt sprouts phosphoric acid, potassium carbonate, trisodium phosphate and ammonium hydroxide are added as nutrients. For acetic acid fermentation, the alcohol content of the fermented mash is adjusted to 7-8 per cent by dilution with water, because acetic acid bacteria do not grow well at higher concentration of alcohol. After this adjustment, mother vinegar containing acetic acid bacteria is added at the rate of one part to ten parts of fermented mash in order to check the growth of undesirable microorganisms and to hasten the fermentation process. The chemical reaction involved is as under:

Fruit extract + Yeast

↓

Alcohol Fermentation

↓

Acetic Acid Fermentation

↓

Ripening

↓

Filteration

↓

Pasteurization

↓

Bottling of Vinegar

Figure 29: Process chart for vinegar

Stage in process: Essential	Stage in process: Optional	Quality Assurance	Equipment Required	Notes
Fermented juice				See Figure 26 for production of wine
	Filter		Coarse muslin cloth.	Filter through coarse musline cloth to obtain juice.
Mix ←	Acetic acid bacteria			Add either a starter culture of bacteria or liquor from a previous batch
Ferment		Check rate of fermentation, check for mould growth and insect contamination.	Fermentation vessel or 'generator'. Filter cloth.	It is important that air is freely circulating in the fermenter and a special design is needed to ensure this.
Filter ←	Bacteria & pulp			
Clear		Check that product is clear.		
Standardize		Check acetic acid content.	Titration equipment.	Adjust the acetic acid content to that declared on the label by blending with previous batches or adding clean water or acetic acid to achieve 6-10% acid.
Fill & Seal ←	Bottles/caps	Check fill-weight and correctly sealed pack.		Cold fill into pre-sterlized bottles.
	Heat	Time and temperature of heating.	Water-bath.	Optional pasteurization for 5-10 minutes at 60°C.
Label & Store		Check that label & storage conditions are correct.		Store on racks, in boxes with a cool dry atmosphere.

Figure 30: Flow chart of vinegar processing

Preparation of vinegar

Vinegar is prepared by the following methods:

Slow process

This process is generally used in India. The fruit juice or sugar solution, filled in earthen pots or wooden barrels, is kept for at least 5-6 months in a warm, damp room to undergo spontaneous alcoholic and acetic fermentations. No special care is taken, but the mouth of the container is covered with cloth to keep out insects, dirt, etc. The main defects of this method are:

- Incomplete alcoholic fermentation;
- Slow acetic fermentation;
- Low yield
- Inferior quality of vinegar.

Orleans Slow Process

The vinegar prepared by this process is clear and of superior quality. The steps of the process are:

1. *Selection of fruit*: Grapes, apples, oranges, mangoes, dates, jamun, or any other sweet fruit of third grade having about 10 per cent sugar in the juice are taken. Cores and peels of certain fruits discarded during canning and jam making can also be used.
2. *Extraction of juice*: The fruits or vegetables are cut into small pieces and then crushed or pressed through a thick muslin cloth. Fruits which do not yield juice readily are heated with a small quantity of water before pressing.
3. *Adjustment of sugar*: Only juice containing low percentage of sugar is suitable for the growth of yeast. The concentration of sugar is determined by means of a hand refractometer and adjusted to about 10 per cent either by diluting the juice with water (if the sugar content is high) or by adding additional sugar.
4. *Fermentation*: The juice is heated (pasteurized) to destroy the microorganisms and then filled in glass carboys, earthen pots or wooden barrels to three fourths of their capacity. The two important steps in the preparation of vinegar are:

   a. *Alcoholic fermentation*: Pure wine yeast, obtained from a winery or a chemist's shop, is well powdered and dissolved in a little warm juice and then added at the rate of 1.5 g per litre to the whole lot of juice with frequent stirring. The mouth of the carboy or barrel is loosely plugged with cotton wool to allow carbon dioxide gas to escape. The gas should be completely removed otherwise it hinders the yeast fermentation. Initially there is continuous frothing which indicates the progress of fermentation, but it ceases after 3 weeks when the fermentation is complete. All the sugar is converted into alcohol as can be seen by testing with a hand - refractometer which indicates 0-1 per cent total soluble solids. During the fermentation the temperature is maintained at 22 to 27^{0}C as fermentation ceases above 41^{0}C and below 7^{0}C. The fermented juice is stored for 1 to 2 weeks for sedimentation and then strained through a cloth, or the clear supernatant is syphoned off into a clean container which is filled up to three-fourths its capacity. Vinegar fermentation should be taken up only after ascertaining that the alcoholic fermentation is complete, otherwise yeast will retard the fermentation. For vinegar fermentation the alcohol content of the fermented liquid is adjusted to 7-8 per cent by dilution with water, because acetic acid bacteria do not grow at a higher concentration of alcohol.

b. *Vinegar (acetic) fermentation*: This is brought about by acetic acid bacteria. Unpasteurized vinegar or "mother" vinegar is added to the product of alcoholic fermentation in the ratio of 1: 10, and mixed well. Thereafter the liquid should not be disturbed otherwise the firm of vinegar bacteria will break and sink to the bottom and consume the nutrients in the liquid without producing vinegar. The mouth of the container is closed with a cork having two holes for proper aeration. The temperature of this liquid is maintained at 21 to 25^{0}C and the fermentation is completed in 10 to 15 weeks when the acetic acid content reaches a maximum. Then the vinegar is syphoned off or strained through a thick cloth leaving at the bottom of the container a turbid liquid, which is used as "mother" vinegar for fresh fermentation.

5. *Aging*: Vinegar prepared by the above method is turbid and does not possess a good taste. It is stored in containers for 4 to 8 months during which the vinegar develops a good aroma and flavour and becomes mellow.
6. *Clarification*: The dear aged liquid should be syphoned out and filtered.
7. *Colouring*: Caramel colour is added for colouring.
8. *Pasteurization*: The vinegar is poured into previously sterilized bottles, corked airtight and the bottles heated (pasteurized) in hot water at 71 to 77^{0}C for 15 to 20 minutes, so that further growth of bacteria is stopped and the strength of vinegar maintained during storage.

Quick process (Generator or German process)

In this process additional oxygen is supplied for the growth of bacteria and the surface of the bacterial culture is also increased resulting in rapid fermentation. The equipment, used known as "Upright Generator", is a cylinder of height 3.66 to 4.2 m and diameter 1.2 to 1.5 m which is divided into three compartments:

(i) Distributing (ii) Central, and (iii) Receiving.

An ideal vinegar should contain only about 0.3 per cent sugar. A higher percentage denotes incomplete fermentation due, to excess of acetic acid during yeast fermentation.

4

Commercial Processing Technology of Selected Fruits and Vegetables for Production of Various Value-Added Processed Products

Modern fruit and vegetable processing industry is by and large, employing sophisticated machinery and equipment and efficient technique of production. There is still a scope for expansion of the industry by developing and exporting value-added products and on more foreign exchange. The choice of products to be manufactured will depend on the readily availability of raw materials in the area where the Processing Unit is supposed to be set up and the consumer preference in the neighbouring market. Preservation of food has to be established in the developing countries for other reasons also such as to generate both rural and urban employment, improve farmers nutrition by allowing them to consume their own processed fruits and vegetables during the off- season, stimulate agricultural production by obtaining marketable products, reduction of imports and meeting of export demand and to develop new value-added products. The food processing industry has now become more of a necessity than being a luxury. It plays an important role in the conservation and better utilisation of fruits and vegetables. To avoid the glut and utilise the surplus during the season, it is necessary to employ modern methods to extend storage life for better distribution and processing techniques to preserve them for utilisation in the off season on both large scale and small scale. Fruit and vegetable processing was first started in an organised manner in 1857, mainly to make pickles and chutneys with a view to meeting the export requirement.

Mango Processing Technologies

Mangoes are processed at two stages of maturity. Green fruit is used to make chutney, pickles, curries and dehydrated products. The green fruit should be freshly picked from the tree. Fruit that is bruised, damaged, or that has prematurely fallen to the ground should not be used. Ripe mangoes are processed as canned and frozen slices, purée, juices, nectar and various dried products. Mangoes are processed into many other products for home use and by cottage industry.

The mango processing presents many problems as far as industrialization and market expansion is concerned. The trees are alternate bearing and the fruit has a short storage life; these factors make it difficult to process the crop in a continuous and regular way. The large number of varieties with their various attributes and deficiencies affects the quality and uniformity of processed products.

The lack of simple, reliable methods for determining the stage of maturity of varieties for processing also affects the quality of the finished products. Many of the processed products require peeled or peeled and sliced fruit. The lack of mechanized equipment for the peeling of ripe mangoes is a serious bottleneck for increasing the production of these products.

A significant problem in developing mechanized equipment is the large number of varieties available and their different sizes and shapes. The cost of processed mango products is also too expensive for the general population in the areas where most mangoes are grown. There is, however, a considerable export potential to developed countries but in these countries the processed mango products must compete with established processed fruits of high quality and relatively low cost.

Figure 31: Fruit blanching machine

Figure 32: Fruit drying machine

Green Mango Processing

- **Pickles**: The optimum stage of maturity should be determined for each variety used to make pickles. There are two classifications of pickles - salt pickles and oil pickles. They are processed from whole and sliced fruit with and without stones. Salt is used in most pickles. The many kinds of pickles vary mainly in the proportions and kinds of spices used in their preparation. One basic recipe for the study of the preparation and storage of pickles in oil is as follows:

Mango pieces	250 g	Turmeric powder	2 to 4 g
Salt	60 g	Fenugreek powder	2 to 4 g
Mustard powder	30 g	Bengal gram powder	2 to 4 g
Chili powder	20 g	Gingelly oil	20 to 30 g

The ingredients are mixed and filled into wide-mouthed bottles of 0.5 kg capacity. Three days later the contents are thoroughly mixed and refilled into the bottles. Extra oil is added to form a 1-2 cm layer over the pickles.

- **Chutney:** The product is prepared from peeled, sliced or grated unripe or semi-ripe fruit by cooking the shredded fruit with salt over medium heat for 5 to 7 minutes, mixed and then sugar, spices and vinegar are added. Cook over moderate heat until the product resembles a thick purée, add remaining ingredients and simmer another 5 min. Cool and preserve in sterilised jars. Spices usually include cumin seeds, ground cloves, cinnamon, chili powder, ginger and nutmeg. Other ingredients such as dried fruits, onions, garlic and nuts may be added.
- **Drying/dehydration**: immature fruit is peeled and sliced for sun-drying. The dried mango slices can be powdered to make a product called *amchoor*. The use of blanching, sulphuring and mechanical dehydration gives a product with better colour, nutrition, storability and fewer microbiological problems.

Ripe Mango Processing

- **Purée:** Mangoes are processed into purée for re- manufacturing into products such as nectar, juice, squash, jam, jelly and dehydrated products. The purée can be preserved by chemical means, or frozen, or canned and stored in barrels. This allows a supply of raw materials during the remainder of the year when fresh mangoes are not available. It also provides a more economical means of storage compared with the cost of storing the finished products, except for those which are dehydrated, and provides for more orderly processing during peak availability of fresh mangoes. Mangoes can be processed into purée from whole or peeled fruit. Because of the time and cost of peeling, this step is best avoided but with some varieties it may be necessary to avoid off-flavours which may be present in the skin. The most common way of removing the skin is hand-peeling with knives but this is time-consuming and expensive. Steam and lye peeling have been accomplished for some varieties. Several methods have been devised to remove the pulp from the fresh ripe mangoes without hand-peeling. A simplified method is as follows: the whole mangoes were exposed to atmospheric steam for 2 to 2 1/2 min in a loosely covered chamber, then transferred to a stainless-steel tank.

The steam-softened skins allowed the fruit to be pulped by a power stirrer fitted with a saw-toothed propeller blade mounted 12.7 to 15.2 cm below a regular propeller blade. The pulp is removed from the seeds by a continuous centrifuge designed for use in passion fruit extraction. The pulp material is then passed through a paddle pulper fitted with a 0.084 cm screen to remove fibre and small pieces of pulp.

Mango purée can be frozen, canned or stored in barrels for later processing. In all these cases, heating is necessary to preserve the quality of the mango purée. In one process, purée is pumped through a plate heat exchanger and heated to 90^0C for 1 min and cooled to 35^0 C before being filled into 30 lb tins with polyethylene liners and frozen at -23.5^0C.

In an other process, pulp is acidified to pH 3.5, pasteurized at 90^0C, and hot-filled into 6 kg high-density bulk polyethylene containers that have been previously sterilised with boiling water. The containers are then sealed and cooled in water. This makes it possible to avoid the high cost of cans. Wooden barrels may be used to store mango pulp in the manufacture of jams and squashes. The pulp is acidified with 0.5 to 1.0 per cent citric acid, heated to boiling, cooled, and SO_2 is added at a level of 1000 to 1500 ppm in the pulp. The pulp is then filled into barrels for future use.

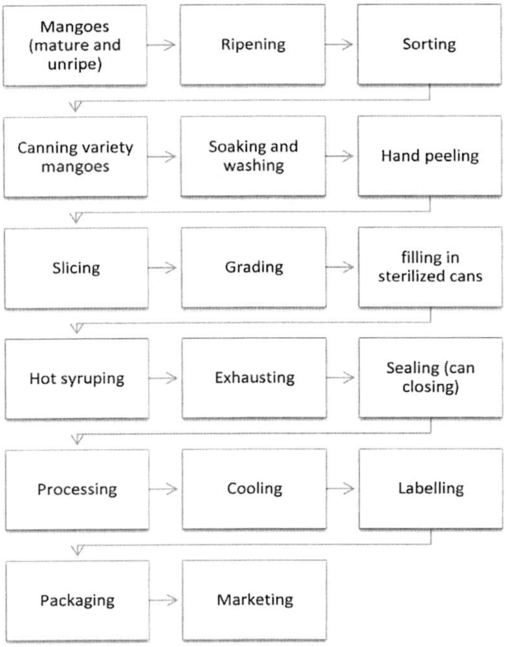

Figure 33: Processing of mango slices

- **Slices:** Mango slices can be preserved by canning or freezing, and recent studies have shown the feasibility of pasteurized-refrigerated and dehydro-canned slices. The quality of the processed product in all of these procedures will be dependent upon selection of a suitable variety along with good processing procedures. Thermal process canning of mango slices in syrup is the most widely used preservation method. Fully mature, unripe mangos are ripened in the cannery to optimum canning ripeness. Mangos high in flavour, with more flesh and low in fibre are always recommended for canning. Sound ripe mangos are soaked in antiseptic and water, brush washed and then conveyed to the preparation tables and hand peeled. "Cheeks" of the peeled mangos are sliced off and longitudinally cut into two or three slices. Side cuts are packed separately. Slices are then conveyed to the filling tables where they are graded for size, colour and maturity and filled into sterile cans. Filled cans are syruped, steam exhausted, sealed, processed in boiling water, cooled, labelled and packaged. Stones with the left-over flesh are steamed and pulped and the pulp thus obtained is packed as such or converted into mango juice or nectar.

Beverages

The commercial beverages are juice, nectar and squash. Mango nectar and juice contain mango purée, sugar, water and citric acid in various proportions depending on local taste, government standards of identity, pH control, and fruit composition of the variety used. Mango squash in addition to the above may contain SO_2 or sodium benzoate as a preservative. Other food grade additives such as ascorbic acid, food colouring, or thickeners may be used in mango beverages.

A short description of finished products found in literature is as follows:

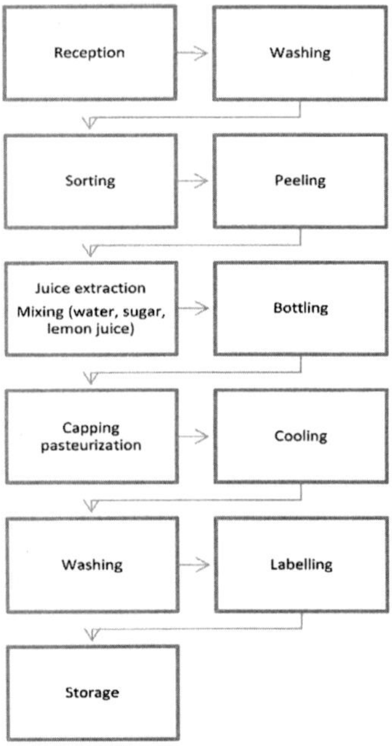

Figure 34: processing of mango juice

Mango juice: prepared by mixing equal quantities of pulp (purée) and water together and adjusting the total soluble solids (TSS) and acidity to taste (12 to 15% TSS and 0.4 to 0.5% acidity as citric acid)

Mango nectar containing 25 per cent purée can be prepared using the following procedure.

	Brix of purée		
Nectar component	15^0	17^0	20^0
Purée	100	100	100
Sugar	45	43	40
Water	255	257	260

Commercial processing conditions may require the use of a preservative. The pH is adjusted to approximately 3.5 by adding citric acid as a 50 per cent solution. The time of heat processing will vary with filling temperature, can size and viscosity of the juice or nectar. Mango squash may be prepared according to flow-sheet described below; the finished product may contain 25 per cent juice, 45 per cent TSS and 1.2 to 1.5 per cent acidity and may be preserved with sulphur dioxide (350ppm) or sodium benzoate (1000 ppm) in glass bottles.

Mango Squash Simplified Flow-sheet

Ingredients

Mango pulp	900	900
Sugar	900	1100
Citric acid	18	15
Water	900	900

Mangoes are washed, stored, peeled with stainless steel knives. The pulp is prepared by using a pulper with fine sieve (0.025-in); Sugar is mixed with water and citric acid = syrup; The pulp is added to the syrup and mixed well; The mixture is strained trough cloth; The squash is heated at 85^0C and bottles are filled and closed. For additional heat treatment bottles may need to be maintained at a product temperature of 80^0C for 30 minutes if the product is to be processed without preservatives. The bottles are then left to cool in water and stored at room temperature.

- Two negative points must be avoided: presence of air bubbles (which is a source of quick deterioration) and separation of squash solids (giving an undesirable appearance). The means to avoid these two phenomena are described in the fruit juices section.

A type of "squash type" beverage may also be manufactured with '/a pulp + '/a water + i/a sugar and pH adjusted to 3.7 by addition of citric acid. Using different sieve sizes affects the quality and reduces air bubbles to a certain extent but homogenization and de- aeration of purée or squash seem to be important in order to avoid separation and air bubbles.

The squash quality is evaluated on the basis of the following characteristics:

- pH,
- Titrable acidity,
- Soluble solids,
- Ascorbic acid (by 2,6 dichlorophenol indophenol method),
- Specific gravity

Processing of Mango Bars

Ripe fruit is used; the mangos are washed, peeled and cut into pieces with a stainless-steel knife. Pulp extraction is carried out with a hand-driven or electrical juice extractor. Sugar, lemon juice (or citric acid) and potassium metabisulphite are then added to the pulp so that mixture contains 25 per cent TSS (total soluble solids) as determined by a refractometer.

The composition of ingredients is as follows:

- sugar: 10 to 15 per cent of the weight of the pulp;
- lemon juice: 2 spoons per Kg of pulp;
- potassium metabisulphite ($K_2S_2O_4$): 2 g per Kg of pulp.

The pulp thus prepared is heated for two minutes at 70-80^0 C. It is then poured into aluminum trays coated with glycerine (this facilitates the removal of the dried pulp). The prepared pulp should be placed on trays at the rate of 15 Kg per square meter. The trays are brought to a sun or solar dryer. The drying is completed when the product has the consistency of leather (about 15% moisture content). The yield ratio between raw material and finished product is about 12:1. Two or three layers of dried product are piled one on top of the other and cut into small squares (4 X 4 cm). Each square is wrapped in cellophane paper, packed in cellophane bags then labelled and stored in a dry place.

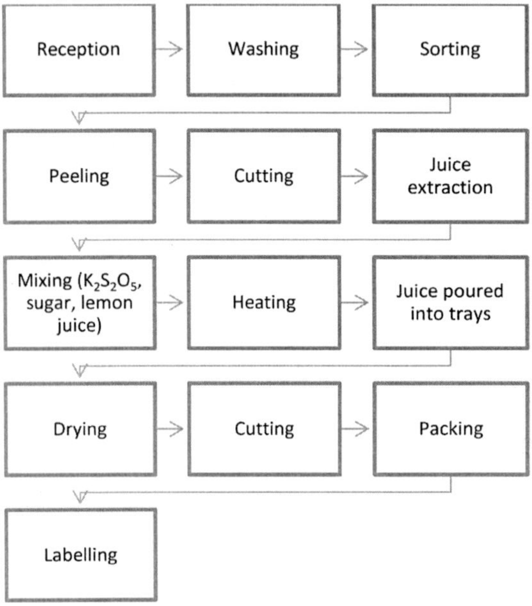

Figure 35: Processing of mango bars

Dried/Dehydrated

Ripe mangoes are dried in the form of pieces, powders, and flakes. Drying procedures such as sun-drying, tunnel dehydration, vacuum-drying, osmotic dehydration may be used. Packaged and stored properly, dried mango products are stable and nutritious.

One described process involves as pre-treatment dipping mango slices for 18 hour (ratio 1:1) in a solution containing 40^{0}Brix sugar, 3000 ppm SO_2, 0.2 per cent ascorbic acid and 1 per cent citric acid; this method is described as producing the best dehydrated product. Drying is described using an electric cabinet through flow dryer operated at 60^{0}C. The product showed no browning after 1 year of storage.

Drum-drying of mango purée is described as an efficient, economical process for producing dried mango powder and flakes. Its major drawback is that the severity of heat preprocessing can produce undesirable cooked flavours and aromas in the dried product. The drum-dried products are also extremely hydroscopic and the use of in-package desiccant is recommended during storage.

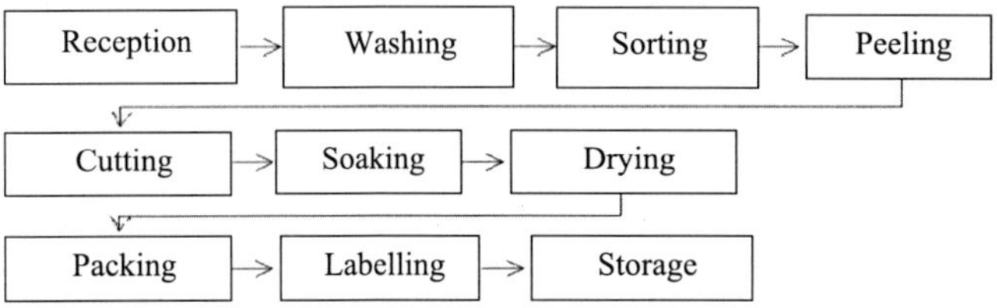

Figure 36: dried mango slices

Guava Processing Technologies Guava Purée

Guava purée is used in the manufacture of guava nectar, various juice drink blends and in the preparation of guava jam. The washed sound fruit is first passed through a chopper or slicer to break up the fruit and this material is fed into a pulper. The pulper will remove the seeds and fibrous pieces of tissue and force the reminder of the product through a perforated stainless-steel screen. The holes in the screen should be between 0.033 and 0.045 in. The machine should be fed at a constant rate to ensure efficient operation.

The puréed material coming from the pulper is next passed through a finisher. The finisher is equipped with a screen containing holes of approximately 0.020 in. The finisher will remove the stone cells from the fruit and provide the optimum consistency to the product. Perhaps the best way to preserve the guava purée is by freezing and the material passing through the finisher can be packaged and frozen with no further treatment. It is not necessary to heat the product to inactivate enzymes or for other purposes. The material can be frozen in a number of types of cartons and cans; however, a fibre box with a plastic bag inside is commonly used and is probably the less expensive.

It is also possible to can and heat process the guava purée and this can be accomplished by heating the purée to 195^{0}F in an open double bottom kettle, filling into cans, closing the cans, inverting the cans for a few seconds, followed by cooling. Cans should be cooled rapidly to approximately 100^0 F before they are cased and stacked into warehouses.

Guava Juice and Concentrate

Guava juice can be used in the manufacture of a clear guava jelly or in various drinks. A clear juice may be prepared from guava purée that is de-pectinised enzymatically. About 0.1per cent pectin- degrading enzyme is mixed into the purée at room temperature; heating of the product at approximately 120^{0}F will greatly speed the action of the enzyme. After 1 hr. clear juice is separated from the red pulp by centrifuging or by pressing in a hydraulic juice press. A batch-type or continuous-flow centrifuge can be used on the de-pectinised purée with no further treatment. The clear juice after centrifuge or after press (and subsequent filtration) can be preserved by freezing or by pasteurization in hermetically sealed cans. For shipment to overseas markets, it may be advantageous to concentrate either the purée or the juice.

Processing of Onions

Drying of Onion

Varieties with pungent flavour are the most appreciated; both coloured and white onions may be used. After removing the tops, roots and outer integuments, onions are washed thoroughly and then cut into slices of 3 mm thick. It is preferable to cut at right angles to the core of the onions. After cutting onion slices are carefully washed. Blanching is not practiced (it makes the onion lose its flavour). The use of preservatives is not necessary; therefore, after washing and draining, the slices are spread evenly on the trays of a dryer. The onions are dried when the ration of prepared raw material to dry product is about 9:1 (moisture content about 5%).

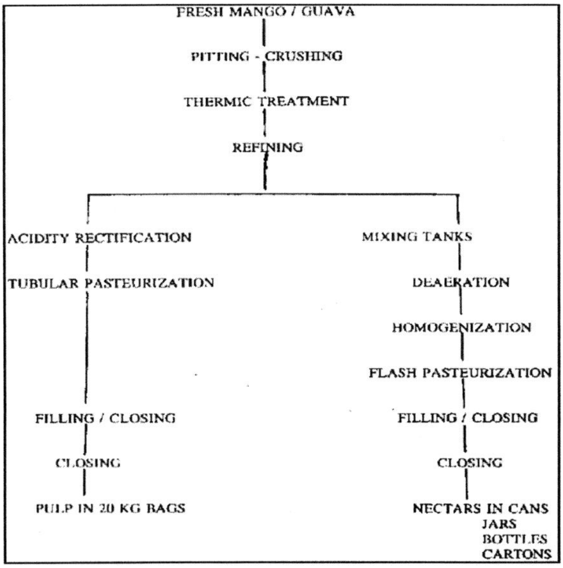

Figure 37: Processing of mango and guava- pulp and nectar processing

Cooling, packing, labelling, and storage follow the same operations as described in other flowsheets. The dried product may be ground into powder (which tend to agglutinate without any anticaking agent). The dryer used for onions must be reserved especially for onion (flavour / odour contamination possible to other products).

Storage life: about 12 months.

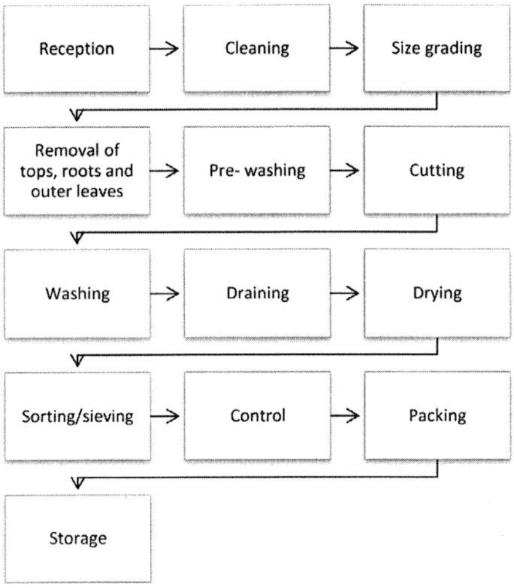

Figure 38: Processing of drying of onions

Processing Technologies for Some Vegetables

- Garlic - The cloves are separated and the outer skins removed by hand. The skins of individual cloves are not removed. The cloves are cut into pieces not more than 5 mm cube. No blanching nor other treatment are applied. The product is spread evenly on drying trays. Dry until brittle. The final moisture content should be about 5 per cent. The yield is about 5:1. After completing drying, separate dry skin by winnowing. The dry product may be ground into powder which should then be protected against moisture pick-up as it is very hydroscopic.
- Beetroot. - After washing, peeling and trimming the material is cut into slices about 2-3 mm thick. Only steam blanching could be used as water blanching will wash out the colour. The slices are immersed in a solution containing 8000 ppm SO_2 for I minute. The prepared material is placed on drying trays and dried to a moisture content of about 6 per cent. Shade drying might also be tried. Overall shrinkage ratio is about 12:1.
- Turnip. - The material is peeled after cutting off tops and roots then cut into slices 2 to 5 mm thick. The slices are submerged for 4 to 6 minutes in boiling water. Slices are dipped for I minute in a solution of metabisulphite containing 8000 ppm sulphur dioxide. After treatments, material is spread thinly and evenly on standard trays and dried to a moisture content of 6 per cent. Shrinkage ratio is 28:1. Turnips are white fleshed, and Swede Turnip yellow or orange fleshed. The latter will produce a dried product of better quality.
- Sweet pepper- The raw material is washed thoroughly and broken up. Cores and interlocular partitions are removed by hand. Defective parts are cut out with knives. The prepared flesh is held under water or a weak solution of sodium metabisulphite (2000 ppm SO_2) before being hand cut into strips about 5 mm wide. The cut pieces are dipped for 1 minute in a solution of 2000 ppm SO_2 (3 g sodium metabisulphite per L). The material is spread in a thin layer on trays and shade dried until crisp or to a moisture content of about 7 per cent. Average shrinkage ratio is about 25:1; drying ratio is about 13:1. For finishing, pick out any burnt or discoloured pieces; put the material on a sieve of suitable mesh to remove small pieces and "fines". Both green and red peppers may be used for drying.

Mushroom and Mushroom Products

Mushroom, a form of fleshy edible fungi is a rich source of high- quality protein, minerals, folic acid and vitamins. Because of pleasant flavour, taste and freshness, mushroom are considered as an important delicacy in human diet. They are praised for their characteristic meaty biting texture and flavour.

White button mushroom (*Agaricus bisporus*), oyster mushroom or *dhingri* (*Pleurotus sajorcaju*) and paddy straw mushroom (*Volvariella volvacea*) are commercially cultivated in India. White button mushroom contributes more than 90 per cent of the total production. However, owing to high moisture content and delicate nature, mushroom is highly perishable and cannot be stored for more than 24 hours at ambient temperature. The deterioration is mainly caused due to their high metabolic activity, respiration rate and susceptibility to enzymatic browning. Drying, canning, and freezing are the accepted methods for preservation of mushrooms and by converting into different products like pickle, chutney, soup etc their availability can be increased.

Canning of Mushroom

White button mushrooms (*Agaricus bisporus*) are preferred over other types of mushrooms for canning. Commercially mushrooms are canned in brine.

Sorting and grading: Diseased, damaged, bruised, shriveled, and browned mushroom are discarded. Only the healthy white and tight buttons are selected and separated into two grades i.e. cap diameter up to 2.5cm with compact head as 'A' and cap diameter beyond 2.5 cm as 'B' grade.

Washing: Graded mushrooms are thoroughly washed 3-4 times in cold running water to remove dirt, soil etc without damaging or rubbing them excessively.

Blanching: To inhibit enzymes activity blanching in necessary. It also removes the air from the tissues of raw material to enable a satisfactory and uniform pack. Mushrooms are blanched in boiling water for 2-3 minutes followed by immediate cooling in water. The steam blanching is preferred because the blanching losses in boiling water have been estimated to be about 30 percent.

Filling of cans: For mushroom canning, lacquered cans (sulphur resistant) are recommended to avoid sulphur staining of canned mushroom. The blanched mushrooms are filled into cans to its declared drained weight i.e. 230 g in A 1 tall can and 430gm in A2½ can size.

Brining: After filling mushroom in cans, add hot brine solution having a ratio of 2 per cent common salt, 1% sugar and 0.05 per cent citric acid. Fill up to the brim of the can. Brining adds flavour to the product, reduces processing time and helps in increasing the shelf-life of canned mushroom.

Exhausting: After brining, cans are exhausted to remove any entrapped air and other occluded gases from the product to ensure longer shelf-life. Cans filled with brine solution are fed to the exhaust box for a specified period depending upon the length of exhaust tunnel and size of can. Exhausting can also be performed by placing the filled cans in boiling water till temperature of the center of can reaches 85-90°C for 1-2 minutes.

Double Seaming /can closing: Immediately after exhausting cans are sealed with the help of double seamer to get the hermetic seam. Sealed cans are then placed in upside down position to sterilize the coded lids.

Processing/sterilization: Processing is also designated as sterilization is an indispensable unit operation in canning. This is accomplished by processing the hermetically sealed cans at a pressure of 15 lbs. psi for a specified period depending upon the size of the can and altitude of processing place. However, for A1 tall and A21/2 size cans, processing is required for 35 and 45 minutes respectively.

Cooling: Cooling of cans is done immediately after sterilization in cold running water to room temperature to give an abrupt shock to the micro-organisms to get rid of their adverse activities.

Mushroom pickle in oil

Ingredient	Quantity
Mushroom	1kg
Salt	100 g
Red chillies Powder	10 g
Turmeric powder	10 g
Coriander powder	10 g
Aniseed and cumin powder	5g (each)
Spices (fenugreek, black pepper, cinnamon powder)	2 g (each)
Garlic chopped/paste	20 g
Onion chopped	100 g
Vinegar	80-100 ml
Mustard oil	200 g

Labelling and storage: The cooled cans are stored in a cool and dry place and smeared with grease to remove any adhering moisture from the can body to avoid rusting. Cans are kept at ambient temperature for 8-10 days to check any swell, leakage, puffing and other disorders before labelling. Before the cans are exposed for sale, proper labelling is done to meet statutory requirements of Indian Food Laws.

Method for mushroom pickling: Wash and cut mushroom into pieces. Blanch by holding the pieces in muslin cloth in boiling water containing salt for 3-4 minutes followed by immersing in cold water. Remove excessive moisture by spreading the pieces on a muslin cloth. Fry onion garlic and ginger in oil, mix other ingredients and mushrooms except vinegar and mix thoroughly. Add vinegar and pack in a clean jar.

Mushroom soup Recipe

Ingredient	Quantity
Mushroom	1kg
Salt	30 g
Red chillies powder, cardamom, fenugreek, black pepper, cinnamon powder	5 g (each)
Butter	100 g
Wheat flour (maida)	12 g (each)
Vinegar	20-30 ml
Milk	200 ml

Mushroom sauce

Ingredient	Quantity
Mushroom	1kg
Salt	20 g
Red chillies powder	10 g
Cardamom powder	5 g
Dry ginger powder	5 g
Garlic paste	2 g
Mace	4 g
Vinegar	100 ml

Preparation of mushroom soup: Take fresh mushroom and wash in water and mash in blender. Pass slurry through an aluminum or plastic sieve to remove the fibrous portion. Melt the butter in a pan and fry the slurry till it is just brown. Add wheat flour, spices and milk to the mixture. Boil the whole mass to a reasonable consistency. After this process, the soup is ready for use. Fill boiling hot soup into the cans and process in a similar way as explained under canning of mushroom.

Preparation of Mushroom Sauce

- Wash the mushroom thoroughly. Keep them in porcelain or an enameled can for 12hr after sprinkling with salt, alternatively dip the mushroom in vinegar.
- Blend the soaked or salted mushroom into fine slurry.
- Add vinegar and spices.
- Heat the mass till it thickens to the consistency of sauce.

- Fill hot in sauce bottles and seal with crown corks.
- Sterilize for 30 minutes at 82-85^{0}C.
- Store in cool and dry place.

Preparation of mushroom chutney: The mushrooms are thoroughly washed in water and cut into small pieces and cooked with a small quantity (100-150ml/kg) of water in a pressure cooker for 15-20 minute. Chopped onion, garlic and ginger are fried in oil and mixed with mushrooms, cook until the whole mass thickness (30-45 minutes). Now add spices and cook for 2-3 minutes. Finally, add sodium benzoate (dissolved in small amount of water) and acetic acid. Final TSS should be in the range of 55-58^{0}Brix. Pack the material in clean glass container while still hot.

Mushroom Chutney Recipe

Ingredient	Quantity
Mushroom Sugar	1kg 250 g
Salt	100 g
Red chillies powder	5 g
Ginger choped	75 g
Aniseed and cumin powder	5 g (each)
Spices (fenugreek, black pepper, cinnamon powder)	2 g (each)
Garlic chopped/paste	25 g (each)
Onion chopped	100 g
Vinegar	10-15 ml
Mustard oil	100 g

Dehydration of mushroom

Material required

Ingredient	Quantity
Mushroom	10kg
Potassium meta-bi-sulphite	150gm
Citric acid	20gm

Method for mushroom dehydration

1. Select sound white button mushroom and wash the fresh produce thoroughly in running cold water and then in salt and 0.1 per cent KMS.
2. Cut into two equal halves longitudinally and wash again quickly in running cold water. If desired these can also be kept as such without

cutting into halves. Keep the mushroom in 2 per cent salt water to check browning.

3. Blanch in boiling water for 4-5 minutes.
4. Cool immediately in cold water.
5. Steep the blanched mushrooms in water containing 1.5 per cent potassium meta-bi-sulphite + 0.2 per cent citric acid for 30 to 45 minutes.
6. Drain the mushroom and wash with water thoroughly.
7. Dry the product by placing over trays.
8. Place the loaded trays in dehydrator maintaining 60^{0}C temperature for 6-8 hours.
9. Pack in polyethylene bag.

Exercises

Module I and II

Answer in short

1. Classification of fruits and vegetables based on moisture content
2. Advantages of food preservation.
3. What is block chain?
4. What is vacuum packaging?
5. What is Argon packaging?

Answer in detail

1. What are the major causes of food deterioration?
2. What are the principles of food preservation?
3. Explain the vegetable retail model.
4. Explain major issues in fruit and vegetable supply chain.
5. Explain minimal processing in brief.
6. Discuss the present trends in the export of processed fruits and vegetables in India.

Module III

Define/ write in short

1. Blanching
2. Microwave blanching
3. Infrared blanching
4. Canning
5. Lacquering
6. Syruping and brining
7. Thermal death rate curve
8. Cordials.

9. Fruit nectar.
10. Preservation by filtration

Answer in brief

1. Explain water bath procedure.
2. What is pressure canner procedure?
3. Thermal death time curve
4. What is exhausting in canning?
5. Objectives of blanching.
6. Preconditions for blanching.
7. Difference between juice and squash
8. Preservation by drying
9. Preservation by sugar
10. Preservation by freezing

Explain in detail

1. Methods of blanching
2. What is hot water blanching?
3. What is steam blanching?
4. Explain ohmic blanching in detail.
5. Factors that influence heat penetration.
6. What is heat sterilization?
7. Significance of enzymes in fruit juice production.
8. What are different pasteurization methods?
9. Preservation with chemicals

Module IV

Write in short

1. What are the basic purposes of food preservation?
2. What is chutney?
3. Preservation with salt.
4. Fermentation with dry salting,
5. Fermentation in brine, or
6. Salting without fermentation.
7. Pickling process.

8. Scum formation
9. Cloudiness
10. Prevention of black neck

Answer in brief

1. What are the principles of preservation?
2. Classification of levels of food perishability.
3. Crystallized fruits.
4. Salient features of jam.
5. Flowchart of preparation of jelly
6. Flowchart of marmalade processing.
7. What is pectin and its importance in jelly making
8. Judging the end-point at which boiling stopped in jam making.
9. What are the common problems in pickle storage?

Answer in detail

1. Explain the concept of value addition in post-harvest management.
2. What are the problems in preparation of preserves and candied fruits
3. Explain the process of making of fruit preserves.
4. Defects in jelly preparation.
5. Explain marmalade processing and its types.
6. Processing of chutney making.
7. What are the factors responsible for good chutney preserve?
8. Explain various pickling processes in detail.
9. What are the causative factors of failure to gel?
10. What are the different types of sauces?

Bibliography

Agricorn.in as retrieved on 20th December 2023. Https://www.agricorn.in/2023/08/value-addition-preservation- food-products.html

AP-42, Compilation of Air Pollutant Emissions Factors from Stationary Sources retrieved as on 16th December 2023 https://www3.epa.gov/ttnchie1/ap42/ch09/final/c9s08-1.pdf

Brian A. Nummer. 2002. Historical Origins of Food Preservation. National center for home food preservation. Https://nchfp.uga.edu/publications/nchfp/factsheets/food_pres_ hist. html

Crop Process Engineering: Lesson 9. Processed Products from Fruits and Vegetables . (n.d.). Http://ecoursesonline.iasri.res.in/ mod/page/view.php?Id=123999

Dauthy ME. 1995. Fruit and vegetable processing. FAO Agricultural services bul[le]tin No.119. Food and Agriculture Organization of the United Nations, Rome, Italy

Directorate of horticulture and food processing, government of Assam. As retrieved on 12th November 2023. Https://dirhorti.assam.gov.in/portlets/post-harvest-processing- value-addition

Food and Agriculture Organization of the United Nations. (1997). au118e.pdf allotment-garden. org/recipe/preserves/fruit-butters- cheeses/ #:~:text=Fruit%20cheeses%20are%20 made%20from,small%20 pieces%2C%20as%20a%20sweetmeat.

Food preservation: jellies, jams and spreads (2024, March 26). NDSU Agriculture. Https:// www.ndsu.edu/agriculture/ extension/publications/food-preservation-jellies-jams-and spreads

Food preservation and storage: sugar preserved food products (n.d.). Http://ecoursesonline. iasri.res.in/mod/page/view.php?Id 111482

Food Technology-I: Lesson 18. Jam, Jelly And Marmalade .(n.d.).Http://ecoursesonline.iasri. res.in/mod/ page/view.php?Id = 147609

Food Technology-I: Lesson 19. Fruit Preserves (n.d.). Http://ecoursesonline.iasri.res.in/mod/ page/view.php?Id =147611

Fruit and vegetable processing - Appendix I - Fruit and vegetable processing flow-sheets . (n.d.). Https://www.fao.org/ 3/v5030e/v5030e0y.htm#Appendix%20I%20-%20 Fruit%20and%20vegetable%20processing%20flow-sheets

Fruit and vegetable processing - Ch08 Fruit specific preservation technologies (cont.) . (n.d.). Https://www.fao.org/ 3/v5030e/v5030e0p.htm

Fruit Preserves and Candied Fruits. (n.d.). Fruit preserves and candied fruits.Https:// gyansanchay.csjmu.ac.in/wp-content/uploads/2022/05/Fruit-Preserves-and-Candied-Fruits.pdf (n.d.).

Fruits and Vegetables value addition by Processing, Dehydration, Canning(n.d.). Https://www.entrepreneurindia. co/project-and-profile-listing?Catid=58&subcatid=58 & catname= Fruits+and+ Vegetables+value+addition+by+ Processing, +Dehydration, +Canning+&+Preservation

FSSR. (n.d.). Technology of fruits and vegetables. In Technology of Chutneys.Https://epgp.inflibnet.ac.in/epgpdata/ uploads/epgp-content/food technology/technology_of_fruits_and_vegetables/21.technology_of_chutneys/et/2851_et_m20.pdf

Https://ugcmoocs.inflibnet.ac.in/assets/uploads/1/134/4620/et/15%20week-4%20 chapter-420030211103030000.pdf as retrieved on 5th December 2023

Lesson 12: general steps in juice processing. Module 4. Fruits and vegetables juice processing. E-Krishi Shiksha. Retrieved as on 24th December 2023 http://ecoursesonline.iasri.res.in/mod/ resource/view.php?Id=147597

Lesson 13: equipment and methods of extraction, clarification and preservation. Module 4. Fruits and vegetables juice processing. E- Krishi Shiksha. Retrieved as on 24th December 2023 http://ecoursesonline.iasri.res.in/mod/resource/view.php?Id=147599

Lesson3 - general principles and methods of food preservation. Processing of horticulture crops e-krishi shiksha. As retrieved on 30th November 2023. Http://ecoursesonline.iasri.res.in/mod/page/view.php?Id=17054

Lesson 41: blanching. Module 5. Food processing equipment and unit operations. e-krishi shiksha. Retrieved on 2 December 2023.http://ecoursesonline.iasri.res.in/mod/page/view.php?Id=3489

National Centre for Home Food Preservation. (n.d.). Https://nchfp.uga.edu/how/ can_06/chutney_principles.html

Noreña, C. P. Z., and Rigon, R. T. 2018. Effect of Blanching on Enzyme Activity and Bioactive Compounds of Blackberry. Brazilian Archives of Biology and Technology 61 Available at: https://doi.org/10.1590/1678-4324-2018180018.

PHT 321: Chutney and sauces (n.d.). Http://ecoursesonline.iasri.res.in/mod/page/view.php?Id=17087

PHT 321: Mushroom and mushroom products(n.d.).Http://ecoursesonline.iasri.res.in/mod/page/view.php?Id =17090

PHT 321: Types of pickles . (n.d.). Http://ecoursesonline.iasri. res.in/mod/page/view.php?Id=17086

Practical manual-food processing and engineering 2023. Integrated Publications, New Delhi, India ISBN: 978-93-5834-006-8 https://egyankosh.ac.in/bitstream/ 123456789/12106/1/Experiment- 1.pdf

Preparation of squash. Daily bloggers blog, posted on August 20, 2015. Https://indiafeelsblog.wordpress.com/2015/08/28/ preparation-of-squash/

Processing of horticultural crops e-Krishi Shiksha retrieved as on 14th November 2023. Http://ecoursesonline.iasri.res.in/mod/ page/view.php?Id=17075

Ranjan S, Dasgupta N., Walia N., Chand C., Chidambaram R. 2016. Microwave Blanching: An Emerging Trend in Food Engineering and its Effects on Capsicum annuum L. Journal of food process engineering. 40.

Reddy, S. C. K., Preethi Sagar, R., & Acharya NG Ranga Agricultural University. (n.d.). Fruit and vegetable processing. In department of food science and technology [Study

Material]. Https://www.rvskvv.net/images/Fruits_Vegetable_Pr ocessing_16.04.20[20]. pdf

Selecting products and production methods www.fao.org/3/W6864E/w6864e08.htm#2.2.5.%20 pectin%20a nd%20papain

T. Ranjani, N. Indra, A. Saranya, Poovizhi Selvi. 2023. Nutrition and Dietetics - Laboratory Manual for Underg[ra]duate Students. SK Research Group of Companies, Tamil Nadu, India.Https://egyankosh.ac.in/bitstream/ 123456789/45130/1/Experiment-7.pdf

W. C. Dietrich, F. E. Lindquist, G. S. Bohart, Herman J. Morris, N. Marvel-dare. 1955. Effect of degree of enzyme inactivation and storage temperature on quality retention in frozen peas. Journal of food science 20(5): 480-491 https://doi.org/10.1111/j.1365-2621.1955.tb16858.x

Wasserman, BP and Wagner JD, 1985. Determination of blanch efficacy by means of a rapid colorimetric peroxidase assay. Biochemistry and molecular biology education. 13(2) pp 49-96 https://doi.org/10.1016/0307-4412(85)90020-2

Xiao, H., Pan, Z., Deng, L., et al. "Recent developments and trends in thermal blanching – A comprehensive review." Information Processing in Agriculture. Available at: https://doi.org/10.1016/j.inpa.2017.02.001.

"Food Preservation and Storage: Canning." Available at: http://ecoursesonline.iasri.res.in/mod/page/view.php?Id=111462.

"Food Technology-I: Lesson 9. Canning and other methods of thermal processing." Available at: http://ecoursesonline.iasri. res.in/mod/resource/view.php?Id=147591.

"So Easy to Preserve", 6th ed. 2014. Bulletin 989, Cooperative Extension Service, The University of Georgia, Athens. Revised by Elizabeth L. Andress. Ph.D. And Judy A. Harrison, Ph.D., Extension Foods Specialists. Retrieved on 1st December 2023. Https://nchfp.uga.edu/how/freeze/blanching.html

Index